The Solutions Are Already Here

The Solutions Are Already Here

Strategies for Ecological Revolution from Below

Peter Gelderloos

First published 2022 by Pluto Press
New Wing, Somerset House, Strand, London WC2R 1LA

www.plutobooks.com

British Library Cataloguing in Publication Data
A catalogue record for this book is available from the British Library

ISBN 978 0 7453 4512 3 Hardback
ISBN 978 0 7453 4511 6 Paperback
ISBN 978 0 7453 4515 4 PDF
ISBN 978 0 7453 4513 0 EPUB

This book is printed on paper suitable for recycling and made from fully managed and sustained forest sources. Logging, pulping and manufacturing processes are expected to conform to the environmental standards of the country of origin.

Typeset by Stanford DTP Services, Northampton, England

Simultaneously printed in the United Kingdom and United States of America

Some repetitions are cliché. Others are priority. I dedicate this book to those we owe a future: to Rowan, to Bruno, to Ara, to Lucía.

Now we push right past to find out
how to win what they all lost

—Santigold

Contents

Preface

This book is about solutions to the crisis that is destroying our planet and so many of its inhabitants. It is not another warning that will only make us more depressed about the problem. Neither is it a quick fix designed to make us feel better for now, as the problem only festers. It is about staring the problem in the face and being honest with ourselves about the changes we have to make to truly solve it.

Because the exploitation of the planet is interrelated with exploitation in human societies, the ecological crisis is very much a question of justice, reparations, or revolution. Consistent with the critiques I present here about climate apartheid and green colonialism, all of the author proceeds from sales of this book will go to Indigenous initiatives against ecocide and for the recovery of their territories in Indonesia and Brazil, to communities who supported this book by giving interviews and arranging for its translation into other languages.

Acknowledgments

I want to send thanks to my editor, Neda Tehrani, for the scrutiny and the ideas on the outline that really helped me understand the book I was writing.

I am extremely grateful to all the people who sat down for interviews, giving this book a global scope and amplifying the voices that are often excluded. Above all I am grateful to the people who put their blood, sweat, and tears into the initiatives and battles covered in the interviews. They are the true creators of the knowledge this book attempts to focus and transmit.

I want give a special thanks to Zenite for the enthusiasm, the help finding amazing sources, and the translations from Portuguese. Thanks to the comrades in Indonesia who made the world smaller, more connected, with their solidarity and dedication. Thanks to Baky for the encouragement and the conversations about one of the more controversial aspects of this book. Thanks to Adrianna for our vagabond dialogue exploring rootedness and interrogating power across three continents.

Thanks to Xander for the good conversations, the support, and all the pdfs and research recommendations that proved invaluable for this book. Thanks to Return Fire for the encouragement and sources. Thanks to T Châu for sources and insight on engaged Buddhism. Thanks to Gabriel Kuhn for generously sharing his book with me. Thanks to Tawinikay for pointers on the Indigenous struggles that provided the necessary context to what is happening today in so-called Canada.

Thanks to Lp for helping me with therapy when I couldn't afford it, and thanks to X and Kevin for financial support. Thanks to M for handing me down that clunker of a laptop from 2006, and to Vayra for helping me spring for a new (used) one.

Thanks to my mom for nurturing in me a love of the natural world and always encouraging me to write, and to my dad for those early walks in the woods to check out what the beavers were up to, and for teaching me how to garden.

Love and fire to all my comrades who have fought alongside me against this insatiable machine, leading the charge or watching my back:

you give me life; to the generations who struggled before us: we carry you with us; and to the land and water that give me sustenance as I write these lines.

Thanks to Gali for the love and cuddles during a winter of pandemic and confinement, and for jumping up in my lap to distract me from writing just the right amount, neither too often nor too seldom. Finally, a sweet thank you to R for making it so hard to finish.

1
A Wide-Angle View

THE BARE BONES: THE SITUATION NOW AND OUR LIKELY FUTURES

Our planet is suffering a crisis that is both catastrophic and unprecedented.

The catastrophe is present all around us. We can measure it, and we can experience it. Even if we begin with a limited focus on global warming, the aspect of the crisis that has received the most attention, we can find plenty of strands that draw our attention to a whole host of other problems that implicate not only how we produce our energy, but also how we feed ourselves, how we are governed, and how we create and share wealth. Following these strands, even in the condensed summary I am about to provide, means dealing with plenty of ugly, depressing facts. Nonetheless, taking in the scope of the problem is necessary for looking at the solutions, and ultimately, that is what this book is about.

As atmospheric carbon dioxide has increased from 250 to 418 parts per million since the nineteenth century, the average surface temperature has gone up by almost 1°C and it is still rising. In a complex system, such a huge change does not mean a smooth, gradual warming, but a major outbreak in turbulence as shock waves ripple all throughout the interconnected systems of the planet. These shock waves include more violent storms,[1] heavier rainfall, more deadly flooding and catastrophic landslides; and on the other hand more intense droughts and widespread wildfires. The west coast of North America, after experiencing its most intense drought in 1,200 years, went up in flames in the summer of 2020, with fire intensity in California and Oregon many times higher than in any year of the preceding two decades.[2] Even the Amazon rainforest is burning.

Increasing temperatures and drought contribute to widespread desertification. When water supplies are disrupted through mining or commercial irrigation and soil is destroyed by deforestation, overgrazing, or commercial monocrop farming, deserts expand. The Gobi Desert is swallowing up over 3,000 km^2 of land every year, and a half a million km^2 of arable land have disappeared in the Sahel in the last fifty years. About 40 percent of the continental US is experiencing desertification, while in Mexico, Paraguay, and Argentina more than half the territory is threatened.[3]

Still other shocks come in the form of deadly heat waves. In temperate and even arctic regions, temperatures have exceeded 40°C for extended periods of time, while new records have been set in Death Valley (54.4°C in 2020) and the Sahara (51.3°C in 2018). Heat waves have increased in frequency by 80 percent due to anthropogenic climate change.[4]

The oceans are acidifying and losing oxygen, threatening nearly all marine species with decline or extinction. Growing swathes of the Arctic are becoming ice free every summer, leading to a loss of habitat and also creating a feedback loop: with less of the planet's surface covered in highly reflective ice, more solar radiation is absorbed, causing even more warming.

The interlinked problems of severe warming, pollution, noxious infrastructures, and extractive industries are causing mass die-offs. One million species are at risk of extinction and animal populations across the board have declined by 68 percent since 1970.[5] Extinctions are currently happening 1,000 times faster than the normal or background rate.

Given that a habitat is a web of mutually beneficial relationships between living species and a host of geological entities such as bodies of water, soil, and air, it is no surprise that entire habitats are disappearing. On a geological timeline, habitats are always changing. Throughout the history of our planet, habitat loss from the perspective of one species is usually habitat gain from the perspective of another species. And though we are right to associate water with life, even the spread of deserts has often been a shift from one kind of biodiversity to another kind.

However, at an accelerating pace over the last century, we have witnessed a wholly different kind of change that can be described as an objective loss of habitat for all living beings: the proliferation of wastelands or dead zones. These are places that, in quantitative terms, have low biodiversity and low biomass. In other words, hardly anything lives there, as though an entire area has been removed from the living world.

A prime example are oceanic dead zones, large areas of an ocean or sea that become depleted in oxygen and subsequently devoid of most forms of life. The dead zones proliferating today are caused by chemicals from industrial agricultural saturating a watershed and causing algae blooms that consume all the oxygen. There are currently over 400 such dead zones worldwide, including in the Chesapeake Bay, off the coast of Louisiana, in the northern Adriatic, the Kattegat strait between the Baltic Sea and the North Sea, and in the coastal waters of China, Japan, and New Zealand.[6]

Another example of a wasteland, a former habitat that our society has made unsuitable for life, are the toxic sites poisoned by a wide variety of industrial practices. Manufacturing—especially in the chemical and electronics industries—mining, and energy production result in huge quantities of toxic waste that is lethal to humans and other life forms. Much of this pollution stays in the environment a very long time, with examples including the radioactive byproducts of nuclear energy, with a half-life of billions of years, or synthetic chemicals like PFOA, a carcinogen used in Teflon that is so stable it is all but indestructible.

These toxins are concentrated at the point of production or intentionally stored in a waste dump. With a cavalier mentality, such sacrifice zones are justified as the necessary price for people to have air fresheners or new cell phones, though in truth no sacrifice zone is perfectly isolated, with carcinogens and other poisons leaking off into the water, soil, or air for the foreseeable future. In other instances, however, poisonous chemicals are intentionally pumped into the environment as widely as possible, as is the case of the 2.5 million tons of pesticides used for industrial agriculture every year.[7]

In the United States, highly contaminated industrial wastelands are placed within the Superfund system, which lists 40,000 toxic sites spread across the country. Fifty percent of the population of New Jersey live within three miles of a Superfund site.[8] Clean-up is paid for by consumers and taxpayers; however, most sites are left to slowly leak out with no clean-up whatsoever.

The impact and meaning of a toxic site are impossible to convey quantitatively. In order to understand just what is being done to the planet, perhaps we need to get a little more visual. The most devastated places I have ever seen were an open pit copper mine in the Atacama Desert and Sierra Minera in Cartagena, Spain.

The Atacama Desert is the driest place on earth. Walking across the face of it feels like being on the surface of another planet. Nonetheless, there are quite a few creatures that live in that seemingly inhospitable place, and the longer you spend there, the more you pay attention, the more you realize how alive it really is, even before you discover the *lomas*, or fog oases that survive by drawing moisture out of the air, and the forests of *chañar*, trees kept alive by groundwater.

The open pit copper mines, operated by multinationals or by the state-owned company Codelco, are nothing like that. The one I saw was like a gaping wound in the earth, too big and brutal to be believed. It was unsettling the way the mine, clearly excavated without any concern for the harm it entailed, was nonetheless dug out in a semblance of geometric perfection—a terraced abyss of concentric rings—like some deeply unhappy creature's idea of beauty. The devastation of the habitat, the scars of heavy machinery, countless tons of explosives, and toxic run-off had resulted in a landscape hostile to life itself. And the death it caused went well beyond the gigantic hole in the ground, nearly a kilometer deep and several kilometers across. All the water stolen by the industry has irrevocably depleted the water table that fragile desert ecosystems depend on. Many once lush forests in desert oases are now graveyards of desiccated trees.

The Sierra Minera of Cartagena has been mined for 2,500 years, since the times of the Phoenicians and Carthaginians. In the mid-twentieth century, multinational mining companies switched to the more profitable open pit mining system. Now it looks like Mordor, which, incidentally, was based on the artillery blasted trenches Tolkien witnessed in World War I, as well as the slag heaps and smoke-choked landscape of the coal-mining and industrial regions of the English Midlands, a comparison that suggests an affinity between total warfare and industrial mining. Denuded hills carved out in unnatural shapes, a long interplay of excavations, the roads flattened to carry the minerals away, and then erosion as mud and rock gave way to wind and rain, and then baked dry in the sun. And everywhere, pools of blood-red goo giving off noxious smells. Countless children in nearby villages are experiencing severe health problems from leftover toxins, years after the mines have been closed.[9]

Next to the toxic sites produced by mining and industry, one of the most common types of wasteland presents quite the contrast. Though they are defining features of landscapes in the Global North, few people would think to include them as examples of a wasteland. In fact, they actually masquerade as symbols of fertility, prosperity, and lush, green bounty in the bourgeois imaginary. I'm talking about the two bookends of capitalist suburbia: green lawns and parking lots. There are over 160,000 km^2 of lawn in the US alone, maintained to the tune of billions of dollars of chemical products, water, and gasoline-powered lawnmowers, making it the number one "crop" in the entire country.[10] This huge expanse, twice as large as all of Ireland, is home to a tiny number of grass species, which are cut short before they can feed any pollinators, and serves as a meager habitat for a small number of bugs. It is, in other words, far more desolate than a desert.

Parking lots and asphalted areas more generally are the companion to the artificially green residential subdivisions. To fulfill their dream of consumer bliss, all those isolated houses with parceled lawns require individualized transportation—cars—and abundant places to leave those cars while shopping and working. (Mortgaged) home ownership, consumerism, and car culture form the normative idea of success and happiness at the center of American capitalism, an idea that has globalized considerably over the past decades. Between roads and parking lots, 158,000 km^2 across the US are covered in pavement. This is almost as much land as is dedicated to wheat farming.[11] In the UK, it's around 8,000 km^2. Aside from constituting a dead zone hostile to nearly all forms of life, parking lots and roads are a source of water pollution and urban heating.

The destruction of the earth's living communities has a major impact on human life as well. One study found that in 2018, one in every five deaths around the world was caused by fossil fuels.[12] The World Health Organization estimates that between 2030 and 2050 climate change will cause an additional 250,000 deaths every year, though this only counts excess deaths (deaths in excess of rates previously considered normal) from more severe heat waves, loss of access to clean water due to climate change, malnutrition caused by drought, and the geographical spread of the malaria zone.[13] The already alarming figure of 2.5 million people killed every decade by the energy, agriculture, and manufacturing industries does not take into account the complex way that different aspects of the ecological crisis are interrelated, beyond just climate.

Take all the deaths caused by contaminated drinking water. Deforestation causes erosion, which, together with the climate trend towards more violent storms, increases flooding, one of the principal ways drinking water is contaminated. And the shift from localized subsistence agriculture to commercial cash crop production (the "Green Revolution" encouraged by leading governments, corporations, and institutions the world over) multiplies the wasteful use of water, as well as poisonous run-off. Contamination of water is also caused by mining, waste dumps, and urbanization. The result is that 500,000 small children are killed every year.[14] While only a small portion of those deaths are directly attributable to global warming, access to clean water is undeniably an ecological issue, a question of how we treat our environment, and what kind of economic activities we promote to "make a living," as inappropriate as that phrase often is.

What about food production? How we feed ourselves is one of the ways we most intensively interact with the rest of the living world. Every year, human societies produce a surplus of food, yet 3.1 million people die from malnutrition and under-nutrition. Even in wealthy countries, millions of poor and racialized people are put at risk of diabetes and heart disease because they live in "food deserts," neighborhoods where it is impossible to obtain healthy, fresh food.

Air pollution, caused largely by cars, energy production, and manufacturing, was already killing 8.8 million people a year in 2015.[15] A study in *The Lancet* found 1.8 million deaths a year caused by water pollution and 1 million deaths a year caused by pollution in the workplace.[16]

Our society produces a tremendous amount of waste, which is bad news for the people—usually poor people—who have to live close to it. Living near a landfill substantially increases the risk of a range of cancers and respiratory diseases.[17] And none of these statistics do justice to the millions of people who are sickened or disabled for life, the people who take care of them, and all the people who have to carry on after losing loved ones.

Because our society is making ever larger areas of the planet unlivable, millions of people are forced to pull up their roots and migrate in search of a more secure place to live. Ecological refugees face the trauma of losing their homes, the racist abuse they endure throughout their migration, and—if they do not join the tens of thousands who die in the Mediterranean or the Sonoran Desert, victims of border regimes that are designed to kill—extreme marginalization when they arrive in the countries that have profited the most off their ecological woes.

In just the first half of 2019, 7 million people were internally displaced (within their home countries) by extreme weather events, which is two times more than the number displaced by violent conflicts.[18] The Intergovernmental Panel on Climate Change (IPCC) has estimated that by 2050, there will be 150 million "environmental migrants" or climate refugees.[19]

In other words, our society's destruction of the earth is very much a suicidal activity and is already one of the greatest causes of death and suffering that humans face.

No one knows what the future will look like, not even the next hundred years. The exercise of modeling likely climate scenarios is problematic because it often obscures the death and destruction that is already taking place. Bandying about different projections of temperature and sea level rise expected by 2050 to decide how urgently we must take action is to implicitly promote the idea that what is going on right now is acceptable, that the present is some gold standard we should try to preserve as closely as possible. The normalization of all of this death and suffering has much to do with who is profiting off the ecological crisis.

It can be useful to guide our efforts to look at the likely changes we may face, but I want to reject any notion of normalizing *only* 10 million human deaths a year or *only* a 10 percent extinction event as some kind of victory.

In the mainstream conversation around climate change, the most optimistic proposal suggests achieving carbon neutral economies by 2050, which supposedly could keep the temperature from rising more than 2°C. What changes could we expect to see in that most optimistic scenario?

The millions of yearly deaths discussed above would increase as clean water becomes scarcer, droughts and extreme weather events multiply, and desertification spreads. Somewhere around 25 percent of species could go extinct.[20] To name just one of the many precious ecosystems that will suffer collapse, 99 percent of coral reefs will die off, leading to the loss of 25 percent of marine species and the livelihoods of 500 million people.[21]

It will be a world rocked by extreme, deadly heat waves breaking all previous records. The land area subjected to extreme summer heat will quadruple.[22] By 2050, the land that 150 million people live on will be reclaimed by the sea, and the land that 300 million people live on will be below the level of annual coastal floods, destroying coastal cities around the world.[23] Further rises in sea level would probably be locked in over the following centuries.

This is by no means a rosy picture. Nonetheless, governments, NGOs, and scientific institutions around the world are banking on this scenario as an acceptable level of collateral damage. It is no wonder that the breathless chorus of mainstream voices cheerleading the optimistic goal of going "carbon neutral by 2050" rarely discuss the extreme suffering and devastation that actually accompany their chosen timeline. City governments around the world run web pages touting their "Smart City" plans for public transportation, ride shares, and green energy. Think tanks and NGOs try to whip up enthusiasm for the few politicians who have actually committed to the goal. And barely any of them mention what that rosy scenario means for the planet and its people.

Yet it's even worse than that. There is no guarantee that going carbon neutral by 2050 will actually function as the meager containment wall it is being sold as. Scientific predictions relating to climate have consistently underestimated the intensity and timeline of projected changes.[24] To name just one example, a summer heatwave in Alaska in 2019 led to a massive salmon die-off. The science director for a local watershed non-profit spoke about a climate model they had prepared just three years earlier, that included moderate and pessimistic scenarios. "2019 exceeded the value we expected for the worst-case scenario in 2069," she told the media.[25]

Runaway warming might be caused by a number of feedback loops that are already reaching their tipping point. When the IPCC first introduced the concept of climate tipping points two decades ago, they believed that no such tipping point would be triggered shy of 5°C of warming. Now they recognize that many tipping points can be triggered with just one or two degrees of warming, and there is in fact evidence that some have already begun.[26] These include the collapse of ice sheets, which would substantially decrease the portion of the earth's surface that reflects solar radiation back into space. As the polar regions warm at an accelerated rate, arctic permafrost is beginning to thaw. This has the potential to release a huge amount of methane, a greenhouse gas roughly thirty times more potent than carbon dioxide. Boreal forests in Siberia and North America are also falling victim to warming through more frequent forest fires and insect plagues. The massive tree and soil die-off means the release of more CO_2.

The Amazon rainforest, currently home to one in ten species on the planet and absorbing 600 million metric tons of carbon a year,[27] is in danger of turning into a giant savanna, or even a desert. Droughts caused by warming, together with deforestation for commercial agriculture, work together to take their toll. The estimate is that when the Amazon loses between 20 and 40 percent of its forest cover, the entire ecosystem will collapse.[28]

Warming in the oceans is causing the slowdown of Atlantic currents that are vital to the transfer of heat and nutrients that form the basis of marine ecosystems, as well as much of the planet's weather. This could exacerbate droughts in Africa's Sahel region and in the Amazon, and would even disrupt the East Asian monsoon, which means the collapse of more habitats, and more suffering for humans and other forms of life.[29]

The implication is that even if we stop all greenhouse gas emissions today, there may be natural processes under way that force a shift to a new dynamic equilibrium, a "hothouse" planet unlike anything nearly all species alive today have evolved to survive.

What might that look like? A 4.5°C rise in temperature could mean 50 percent of species would go extinct, and that's only in a short-term analysis.[30] By the end of the century, 1 billion people would be displaced and hundreds of millions would fall victim to famine. Fifty-five percent of the world's human population would suffer more than 20 days of lethal heat a year; it's more than a hundred days a year in the middle latitudes. Between scorching conditions and the collapse of insect populations, crop yields could decrease by a fifth or more.[31] It's no wonder that even the World Bank says that 4°C of warming might be "beyond adaptation" for human civilization.[32] The hot period could easily last 200,000 years.[33]

As we shall see, the experts cannot solve this problem, and they have already wasted valuable decades. The subtext to the official conversation belies a staggering apathy. *We will not be the ones to die. All those who disappear, human and otherwise, are an acceptable loss. We will come out on top.*

For many people—especially among policy makers and experts—there is a truth to that mindset, at least for now. The millions of human deaths caused by the ecological crisis every year are not shared equally. Most of them occur in the Global South.

However, while the semantic distinction between Global North and Global South is useful, many of the same processes occur in both places; the world is not as divided as those on top want to believe. For example, though the 60,000 people killed on average every year by extreme weather events mostly live in the Global South, so-called wealthy countries are not immune. The 2003 heat wave in Europe, for example, led to 70,000 excess deaths. Needless to say, few of them were living in the houses of the wealthy, with their high ceilings and air conditioning. And while 92 percent of pollution-related deaths occur in low- and middle-income countries, 800,000 people die every year from air pollution in Europe and 155,000 die every year in the US.[34] Still, even these deaths are unevenly distributed. Not many rich people live near industrial parks and toxic waste dumps.

In settler states like the US, Canada, Australia, and Argentina, class is largely inscribed by the historical legacy of colonialism, with the descendants of enslaved Africans and Indigenous peoples subjected to conditions that the global distribution of wealth and power usually reserves for the Global South. When Hurricane Katrina descended on New Orleans in 2005, killing 1,800 people, anyone paying attention saw that the way infrastructure was built in poor and Black neighborhoods left people vulnerable, whereas infrastructure in wealthy white neighborhoods was designed to protect people. And contrary to the spontaneous initiatives of mutual aid that constituted the primary life saver, with neighbors helping neighbors survive the storm, and ex-Black Panthers and anarchists setting up the first on-site clinic,[35] government responses focused on shooting neighbors trying to take clean water or diapers from supermarkets, and then making sure that only middle-class and wealthy residents could return to the city, "gentrification by God." As Neil Smith wrote in the aftermath of that storm, "there is no such thing as a natural disaster."[36] The disaster was produced and directed by economic and political structures.

Those who currently hold power in our society, those who have failed us tragically, do not have our interests at heart, nor those of the planet. And in fact, our interests and the interests of the earth are one and the same. We do not know how disastrous these next decades will be. But there is one certainty that can give us hope and courage: there is not a single scenario in which taking action, in defense of ourselves, in defense of one another, in defense of all the interconnected life on this planet, will not make things better.

IN THE BIOSPHERE EVERYTHING IS CONNECTED: THE ECOLOGICAL CRISIS BEYOND CARBON

The default assumption in our society has been that nature is mechanical rather than communicative. For something to be communicative, it has to have subjectivity, and if it has subjectivity it becomes harder to justify treating it like our personal toilet or gold mine.

Although there have been extraordinary biologists and other experts who have seen in the living world the same mutuality and communicativity that others see, the history of the scientific method from Francis Bacon to the present has in other ways been a process of trained scientists getting dragged kicking and screaming, sometimes by their own research, away from the notion that intelligence or personhood are the exclusive properties of educated white men. Nonetheless, that original assumption still frequently pops up as the default in many intellectual circles.

It feels false to state what has become undeniable: that we can benefit a great deal from observing and learning from other forms of life. The reason it feels so is that to pass from a society that treats the rest of the world so horribly to one that sees "benefit" in learning from others is just a continuation of extractivist models of knowledge that are part and parcel of the extractivist economy that has wreaked havoc on the earth's living communities.

Recognizing that other living beings have their own voices helps us perceive the ways everything is connected, which in turn shows that we cannot take a piecemeal approach to the ecological crisis.

A few years ago, I started keeping a list of all the bird species I saw in my area. Anyone who does birdwatching knows that, even if you appreciated birds before, you really begin to observe them on a different level once you train your gaze to distinguish between one species and another. Few species stay still long enough, in plain view, with good lighting, to allow for a positive identification on size and color alone, assuming you do not have an expensive pair of binoculars. Instead, you have to look at their beak (nutcracker, flesh-eater, or insectivore?), what kind of tree they're hanging out in, group behavior, flight patterns, and songs.

Even if I had previously loved birds in the abstract, I had no specific relation with them and so they were, for all intents and purposes, scenery. The moment I began treating them as beings with personhood and a specific place in a web of relationships, a whole world opened up, and I was enriched by the complexities around me.

One of the most striking things I noticed is that birds know when spring arrives. Since I began watching them, there comes a day early in the year, when from one day to the next, a dozen or more species change their behavior dramatically. Where they had spent the winter extremely human-shy and often hiding, now they are looking for the best places to stand in the sunshine and sing with utter abandon. Moreover, their behavior that day is different from the courtship rituals that would begin a little later in the year, and included birds that already had mates. The extreme difference in behavior left a quantitative register as well: the first year I noticed it, I was able to positively identify as many species in a day as I had during the course of the preceding winter. The only way to describe it is that they were celebrating.

Proponents of the default assumptions of Western thinking will make the unsurprising claim that this is anthropomorphizing, projecting human characteristics on non-humans. Such an unfortunate coincidence that we have no term for the inverse flaw, assuming that only humans possess what are actually widespread traits. In recent decades, biologists have rediscovered what others never forgot: that other living beings think, feel, learn, play, and can be sad or happy. Ritual, culture, intergenerational learning, and mourning are also being documented in a growing body of research, so we may as well get ahead of the curve and speak frankly about celebration, too.[37]

It gave me great joy to discover this sudden change, shared across multiple species of birds by some unknown consensus. And I mean discovery not in the sense of knowledge that I produced, but knowledge that was shared with me when I had the humility or good sense to respectfully approach another community of living beings and see what they had to teach me. That joy was a sort of non-instrumental knowledge that for me was the most important thing, even though it is a type of knowledge our society places a low premium on. It was accompanied by instrumental forms of knowledge as well.

For example, the birds' declaration of spring was not merely a subjective, culturally inscribed proclamation. Their affirmation also has about it something of the *cold, hard fact*. Every year so far, after the day marking their distinct change in behavior, the temperature has gone up and the nightly frosts have ended. The fact that the birds are making a weather prediction with at least some degree of accuracy, and freely sharing this knowledge with anyone who cares for it, is relevant to me because I keep a garden. If I plant my tomatoes before the last frost, no more tomatoes.

And this knowledge takes on a new level of significance as we follow it through time. So far, the day has tended to come earlier year after year. In 2020, it came a half month earlier than 2019, with the birds already conducting their ostensible celebration in mid-January. When we pay attention to the world around us, we can see the signs of climate change, and a great deal more.

I live in Catalunya. Starting in March, we experienced almost two months of strict lockdown. These conditions led to a remarkable improvement in air quality due to the mass reduction in chemical and noise pollution. After all, car traffic had come to an almost absolute halt.

While traffic and air quality returned to their abysmal normal by the summer, there were several effects that lasted throughout the entire year. The change in insect populations was the most dramatic. Just counting the wave of moths that came into my apartment every night during the summer, the number of species increased by a factor of five and the total volume by a factor of two or three. And by paying attention to it in this way, the intrusion became a cause for celebration, rather than for the purchase of some chemical insecticide that would invariably end up in my own drinking water. The number and variety of spiders in the fields beyond my neighborhood also increased noticeably, and while spiders are not insects, the species they depend on for food are.

The catastrophic collapse of insect populations is a cornerstone of the ongoing ecological crisis.[38] Insects occupy several vital nodes in the food web, aiding with the decomposition of plant and animal matter, helping turn dead creatures into the soil that forms a basis of most living cycles; serving as the primary pollinators for nearly all flowering plant species; and serving as a prime food source for small birds, reptiles, amphibians, and mammals, and thus, indirectly, as a food source for the large birds, reptiles, amphibians, and mammals that eat those. The loss of insects threatens all those other species. Without them, the webs of life that currently populate the Earth become impossible to imagine. Insect populations are falling victim to a diverse array of threats, including rising temperatures, water and air pollution from cars and manufacturing, and large-scale pesticide usage. This corner of the catastrophe is a key example of why it does not make sense to compartmentalize the ecological crisis or focus reductively on climate. It is, most certainly, *all connected.*

And we are connected with it. Localized knowledge of the kind I am describing, though it can be brushed off as anecdotal, is important because it accentuates our consciousness of these problems, it motivates us to care for the other beings that are affected and thus to take action, and it can also help guide that action. My observations about the effect of car traffic on insect populations are not enough to confidently claim that getting rid of cars will solve the problem, but it is certainly enough to suggest that putting a stop to car traffic might well be a good start. The localized decrease in pollution was connected to a quick rebound, in at least some species, that was robust enough to continue even after pollution returned to normal levels.

When everyone is sensitized to the other forms of life we share this planet with, we will be more knowledgeable, more agile, and more rapid in responding to threats to the ecology, more able to heal our ecosystem. And our territorial knowledge can be complemented by scientific research, accelerating processes of empirical understanding and reducing the burden on specialized nodes of scientific production that have been too easily bought off by the very industries that are killing us, leaving conscientious scientists actually trying to fix the problem as an overextended minority.

Learning that it is all connected also means learning to connect. Ornithologists and conservationists trying to restore devastated Atlantic puffin populations on the islands of coastal Maine failed year after year until they learned "to think like a puffin."[39] Atlantic puffins were decimated by commercial hunters in the late nineteenth century. By the late twentieth century, the islands they might use as breeding colonies were either wildlife refuges protected by the government, or simply isolated rocks no longer targeted by any extractive industry, hunting or otherwise. Though they were safe, puffins did not return to breed on many of the small islands where their populations had been extinguished, even after conservationists started hatching eggs on those islands. It was only when conservationists started thinking about how puffins are intensely social creatures that they started to change their tactics.

After going out to sea for the extended hunting season, what would cause a returning adult puffin to decide not to start nesting on the island of their birth? Maybe they would be "too timid" to start nesting on a rocky islet on which there was no sign of other puffins? The conservationists began making puffin decoys and placing them across the islands that were now a safe breeding territory for the bird. The technique was a success, and led to Atlantic puffins recolonizing a significant part of their former range. Since then, similar techniques have been used to aid 42 seabird populations in 14 countries.

Sometimes, reintroducing a missing species creates a trophic cascade that changes an entire ecosystem in a way that makes it more robust, more able to withstand climate change and other aspects of the ecological crisis. Sea otters have been involved in several such trophic cascades across the west coast of North America. The otters were hunted almost to extinction by commercial fur traders in the eighteenth and nineteenth centuries, disappearing from most of their range. Since then, they have made a recovery in several areas, sometimes on their own, sometimes with the help of conservationists, and the effect has been remarkable. In the Aleutian Islands of Alaska, they brought down sea urchin populations, which had exploded in their absence. With fewer urchins, the kelp forests could thrive again. And in several California estuaries, they have contributed to a growth in sea grass of 600 percent in just three decades, in part by eating the crabs that eat the sea slugs that keep sea grass clean and healthy. This is remarkable for several reasons. Conservationists were scarcely aware that estuaries even constituted a part of sea otter's natural habitat, and the US government left estuaries out of the otter recovery plan. Furthermore, sea grass beds should not have been recovering in the estuaries feeding into Monterey Bay and the San Francisco Bay because of how polluted these water systems are by chemical and nutrient run-off from California's massive, for-profit agricultural industry, showing how much difference a healthy web of relationships can make in mitigating the effects of pollution.[40]

Sea grass and the salt marshes that are also starting to recover with the aid of sea otters are so important because they help decrease coastal erosion during storms—a problem that will only grow through the twenty-first century—and they constitute an important carbon sink, a habitat that takes carbon out of the atmosphere. They are also an important protective habitat and source of food for numerous other species. The same is true of the kelp forests. "The difference in annual absorption of atmospheric carbon from kelp photosynthesis between a world with and a world without sea otters is somewhere between 13 and 43 billion kg (13 and 43 teragrams) of carbon" according to marine biologist James Estes.[41]

Moving from Monterey Bay to the driest place on earth, we find another example of interconnectivity. The *lomas* are fog oases in the Atacama of northern Chile and the coastal deserts of Peru, a region where rainfall is almost completely absent, measured in mere millimeters a year. Despite the lack of precipitation, diverse ecosystems have been able to thrive thanks to the mist that frequently appears. Taller trees in the *lomas* have specially evolved to condense liquid water from the fog, capture it, and guide it down to the roots, sharing it with all the other plant species that make a life there. Despite the desert conditions, there are 1,400 plant species throughout the different *lomas* forests, nearly half of them found nowhere else.[42]

Most of the *lomas*, however, have been damaged or destroyed as a result of Spanish colonization and the exploitative economic practices that it unleashed, such as commercial grazing, logging, and mining. While they once covered 15,000 km^2, they have now shrunk by 90 percent of their former range.[43] The city of Copiapó, called "Saint Francis of the Forest" by the Spanish, now sits in a desolate valley mostly devoid of trees, and even its river has dried up. Yet humans were a healthy part of *lomas* ecosystems. Before colonization, the Chiribaya culture irrigated 85 hectares of farmland and had grazing land for alpacas and llamas at a town near Ilo, fed entirely by four springs coming from the *lomas*. Now the area is mostly denuded.

Conservation efforts in the *lomas* initially did not take into account the role Indigenous peoples have played in their ecosystems. Ecologists rediscovered what the land's Indigenous inhabitants had already known: that certain taller trees like *Caesalpinia spinosa*, the tara tree, played a keystone role in the *lomas* ecosystem by capturing water from the fog. Yet when they tried and failed to reseed a new generation of tara trees, their investigation into why reseeding efforts were unsuccessful finally revealed what should have been the most obvious oversight. Genetic analysis suggested that Indigenous peoples intentionally seeded and cultivated tara trees, aided by their llamas and alpacas, who ate the fruits and digested the coating of the seeds, making them more likely to germinate. The cattle, sheep, and goats favored by commercial/colonial agriculture, incidentally, will not touch the tara fruit.[44]

A similar theme begins to emerge when we look back at the catastrophic pattern of wildfires in western North America. About three years ago, it began to be widely reported in mainstream media that Western fire suppression techniques were largely responsible for forests becoming saturated with explosive amounts of brush that lead to cataclysmic wildfires, and that the traditional techniques of Indigenous peoples from the region like the Karuk, Yurok, and Mono were actually far superior. They constituted a folk technology designed to prevent cataclysmic wildfires as well as to ensure a healthy forest with high biodiversity and plenty of species humans could subsist off of.

With a satisfied tone that suggested some historical wrong had been corrected, many of these articles mentioned that government forestry and parks bureaucracies would begin to consult with the Indigenous former custodians of the land. Or they simply mentioned that scientists had begun to "study" Indigenous techniques, an ambiguity that is ominous to anyone aware of what it has historically meant for scientists to study non-Western peoples.

None of the mainstream articles I could find mentioned the especial cruelty with which white people occupied the western part of the continent, from the Spanish and Mexican concentration camps to the massacres of the California gold rush. None of them offer much detail on how Native controlled burnings were part of a web of interrelated practices—both economic and spiritual—that linked field agriculture with forest gardens with hunting, protecting an ecosystem that offered humans a healthy and varied diet and ensured a place for a diversity of other species, and that gave people meaning as a part of that ecosystem, in which the other living beings were their relatives. Few of them even mention how Indigenous forestry techniques were suppressed together with their language, their religion, their practice of commoning rather than private property, all while they were being violently forced off their land. And none of them discuss where those Indigenous nations are today, what sorts of economic conditions and social discriminations they face, and how they are fighting for justice or redress. At most they describe government–tribe "partnerships" apparently devoid of conflict or inequality. Even alternative, progressive media tended to be weak on these details.

After the major, catastrophic wildfires in Australia in 2019 and 2020, similar articles began to appear with regard to Aboriginal fire and forestry techniques, replicating all the same patterns. A typical mainstream news article is full of quotes from settler academics, but does not quote nor mention by name a single Aboriginal person, community, or people.[45]

As Métis and Cree writer Mike Gouldhawke remarked, "Acknowledgment isn't the opposite of erasure, doesn't stop erasure, & can facilitate further erasure & things worse than erasure. Colonizers have been playing the two off each other since the beginning of colonialism."[46]

What becomes undeniable the more we look is that all across the world humans are a fundamental part of the ecosystems we live in, from the Arctic tundra to the temperate forests to the Amazon and the Atacama. But it is also hard to deny that there are very different manners of being human, with a huge gulf between them. There are the humans who constitute a keystone species, consciously holding their local ecosystem together with intelligent, traditional, decentralized, and adaptable technologies, and there are the humans who constitute the number one threat to life on this planet.

The idea that humans are the bad guys in the ecological crisis is too simplistic. It lets us off the hook for the soul-searching we need to do to fix the problem. Because if we are the bad guys, there is something inevitable about what is going on, and the best we can hope for is to choose the lesser of evils. This plays directly into conventional conservationism: preserving some small parcel of nature—roped off from our evil influence, of course—to mitigate the devastation and also to visit occasionally, paying an entrance fee perhaps, and to reflect on some mythical Paradise Lost.

When we begin to look at connections between the destruction of nature and the destruction of other humans, the narrative becomes more complex.

The looming extinction of the orcas of the southern Salish Sea has garnered a good deal of media attention, focused especially on one Puget Sound pod. No whale in that pod had given birth in three years, until one female got pregnant in 2018. However, her baby died soon after birth, and she carried the body around in mourning for 17 days, over more than 1,000 miles. The orcas have all but stopped breeding for a variety of reasons that are not fully understood, but all of which have to do, not with humans *per se*, as the Puget Sound orcas have happily been neighbors to humans for thousands of years, but with this second type of human whose outline is starting to emerge. Principal causes likely include noise from the many large ships that ply the waterway, as well as declining chinook salmon populations, victim to the hydroelectric dams that clog up most of the rivers in the watershed.

Industrial pollution is another factor. In the five years up to 2018, Boeing dumped thousands of times over the legal limit of extremely toxic PCBs into the Duwamish River, part of the Puget Sound watershed.[47] A leader in the aerospace industry, Boeing is a major armaments producer. Is there a relation between Boeing's habit of releasing poisonous chemicals into the environment with no regard for the harm they will cause, and their willingness to profit off machines designed to kill, destroy, and maim, primarily for the benefit of the country most likely to use weapons against civilian populations in other parts of the world? To suggest otherwise seems willfully naïve.

A different war measure subsidized by the US government in the nineteenth century presents an unlikely precedent to Boeing's amoral profiteering, a war measure that completely transformed the Great Plains of North America. Throughout that century, the settler government pursued a series of wars against hundreds of Indigenous nations as part of its design to capture an entire swath of the continent, "from sea to shining sea." The idea of "Manifest Destiny" synthesized Puritan notions of white supremacy with the equally supremacist and teleological scientific ideas of the day, from the clockmaker god of Enlightenment "Founding Fathers" to the social Darwinism of the late nineteenth century.

Indigenous nations that successfully fought back against the US war of conquest belied the supposed destiny of continental domination and threw a wrench in the gears of Euro-American progress. In particular, the alliance of Lakota, Cheyenne, and other Great Plains nations, the victors of "Red Cloud's War" of 1868 and the vanquishers of Custer's Seventh Cavalry, blocked a major conduit in the colonization of the West and thwarted Yankee pretensions of superiority.

US war planners came to realize that Indigenous knowledge of the territory and their ecological niche as nomadic hunters sustained by the vast herds of millions of bison gave them advantages in guerrilla warfare that would be hard to overcome. So they began a program of offering bounties for the killing of bison, depriving their enemies of their primary food source and imposing dependence on the reservation system, a euphemism for the concentration camps that served to subject imprisoned nations to police control, suppress Indigenous languages and spiritual systems, and impose a Western socioeconomic order based on individualized private property, field agriculture, and the consumption of addictive commodities. General Sherman, who had applied total war techniques in the Civil War, destroying the South's food supply, organized military expeditions and supplied private hunters to eradicate the Native food supply. Lieutenant General John Schofield was transparent in his memoirs: "With my cavalry and carbined artillery …, I wanted no other occupation in life than to ward off the savages and kill off his food until there should no longer be an Indian frontier in our beautiful country."[48]

As bison herds dwindled from 100 million to just a few hundred in under a century, an added benefit accrued to the settler state. With the enclosure of lands, the suppression of Indigenous resistance, and the near extinction of the bison, coupled with murderous union-busting tactics also being deployed in those years, there were no more impediments to the construction of the transcontinental railroads vital to the occupation and industrialization of the west coast, a prerequisite on multiple levels for the likes of Boeing to pollute watersheds and build war machines a century later.

The replacement of bison with for-market field agriculture and cattle herds belonging to individual owners who sent their livestock by railroad to the industrial slaughterhouses of Chicago and thence to feed the blooming populations of the east coast metropolises, directly contributed to one of the greatest ecological disasters of the early twentieth century, the Dust Bowl.[49] During the years of the Dust Bowl, over 400,000 km^2 of soil dried up and blew away, incidentally allowing the banks to appropriate a huge swath of territory by foreclosing on small farmers.

What we have here, then, is a strategic war measure ideated, organized, and subsidized by the state, deployed by settler paramilitary-entrepreneurs in that pattern of diffuse police action particular to settler democracies, that fed into capitalist projects of industrial development and primitive accumulation, which in turn simultaneously operated as both cause and consequence of ecocide. In other words, a broad expanse of thriving ecosystems was intentionally destroyed as a counterinsurgency strategy of *catching the fish by draining the pond*. The continued destruction of those ecosystems was accelerated by the capitalist economic processes that the counterinsurgency measures deployed by a colonial regime had made possible.

In broad terms, this constant counterinsurgency is being waged not against an inert nature, but against a dynamic practice: commoning. Commoning is the practice of holding the land in common, but it also goes much deeper than that. It can be viewed as an ecological rather than an economic practice, as the goal is not to alienate wealth, to produce commodities for a market and thus to accumulate capital that can be reinvested in more ecologically destructive/economically productive activities. On the contrary, the goal is to procure the health and well-being of the members of the community in a way that preserves their autonomy, through an interdependence with their local ecosystem rather than a forced dependence on an accumulative economy. This means that commoning constitutes a roadblock to capitalism and also a zone of illegibility and resistance to state authority. Due to their ecological interdependence, commoning societies tend towards ecocentric worldviews and practices, though this does not mean they are immune to the world-hating philosophies present in Christianity and industrial scientific progressivism that often serve as the handmaidens to colonialism.

The war on the commons has in fact been one of the main philosophical and economic endeavors of liberalism from the eighteenth century to today. One aspect of this war has been the now discredited myth of the "Tragedy of the Commons," an idea put out by major landowners with no direct experience of commoning who wanted very much to warn their fellow man that preventing landowners from privatizing common resources leads to scarcity and chaos. Of course, it leads to a scarcity of investment opportunities for those who would own all of society, and by "chaos" the ruling class has always meant the self-organization of the lower classes.

The commons have been suppressed not by reason but by centuries-long military and police campaigns that continue to this day.[50] The British government, both at home and in their colonies, frequently employed the death penalty against peasants making use of communal resources according to traditional norms,[51] and a large part of the substance of colonization was the destruction of Indigenous practices of commoning in Africa, Asia, and the Americas.

In the Spanish state, areas with long traditions of defending the commons were at the forefront of the anarchist collectivizations during the Civil War in 1936. One of the responses of the Franco regime was to accelerate the depopulation of the countryside, in part by flooding several mountain valleys for the construction of major hydroelectric dams. Writers like James C. Scott have already remarked on the tendency of dictators to favor huge infrastructural projects, especially when these destroy ecosystems that had long served as protective habitats for resistance movements, like the Pontine Marshes near Rome or the Mesopotamian Marshes in Iraq.[52]

The depopulation of the Pyrenees had already begun in earnest in the nineteenth century through deforestation to provide fuel for the iron works necessary to industrialism and the building of the railroads.[53] The railroads were a convenient vehicle for directly seizing and enclosing communal lands, as well as being necessary for the economic acceleration of industrialization. Deforestation, meanwhile, led to erosion and ecosystem collapse which directly weakened the position of rural commoners. Commoning tends to encourage and rely on a diverse ecosystem, contrary to the monocrop deserts of commodity agriculture. With erosion and deforestation in the mountains, pastoralists and the *trementaires*—wise women who traveled between mountain communities collecting herbs and practicing popular healthcare—were among the most affected. Together, they constituted some of the most ardent defenders of the commons as well as an autonomous system of knowledge, movement, and sustenance. With the loss of grazing land and plant biodiversity, they had little choice but to look for jobs in the mines, the industrial colonies, and the cities, though their rebellious spirit would continue to break out in those artificial environments over the next decades.

Over a long timeline, we see a familiar pattern: an ecological collapse that is the deliberate result of a state strategy for counterinsurgency as well as a consequence of the economic expansion that accompanies the expansion of state power. The state appears as the prime strategic agent of these changes, yet very different types of state promote the conditions of capitalist accumulation that the owning class require.

In order to understand this dynamic, we can turn to a concept that emerged from the revolutionary struggles of the nineteenth century and that has subsequently been utilized in the framework of insurrectionary anarchism: *social war*.

André Léo developed the concept of "social war" not as a war of "everyone against everyone" or against nature, but of the state and the bourgeoisie against society.[54] Léo was an anti-authoritarian feminist, survivor of the Paris Commune, and victim of Marx's purge of the First International. Léo was a participant in the 1871 Paris uprising, which constituted a revolt against the French monarchy and also an important experience in self-organization for the growing workers' movement. The Commune was drowned in the blood of up to 20,000 communards executed in its aftermath. A witness to how the repression was instituted in a more permanent form after the firing squads had finished their gruesome labors, and was worked invisibly into the very fabric of society's political and economic forms, Leó theorized that the statist, capitalist order constituted a constant war of the dominant classes against all of society.

Insurrectionary anarchists like Alfredo Bonanno used *social war* as an expansion on *class war* to deploy an analysis of how all of society—even the organizations and ideas of the workers' movement—was structured to pacify and recuperate resistance and perpetuate the endless parade of alienation and commodification.[55]

Current understandings of *social war* incorporate an awareness of counterinsurgency, the scientific methodology of state planners to understand society as being in a permanent condition of resistance against established authority, and the consequent need to eternally intervene to recuperate, institutionalize, or otherwise surveille and pacify rebellious populations to prevent their resistance from becoming more combative and insurrectionary. Counterinsurgency methodology was largely developed by colonizers in Kenya, Algeria, and Ireland, and quickly exported across the world to theaters of conflict like Los Angeles and Vietnam.[56] In recent years, anarchist researcher Alexander Dunlap has used the lens of social war to describe state interventions in imposing and protecting extractivist infrastructure in colonized territories from Oaxaca to the Andes.[57]

We can place this conceptualization in dialogue with the model for the modern capitalist state proposed by systems theorists like Giovanni Arrighi: the marriage between a political-military class focused on the domination of a *space of places*—territorial conquest, in simplified terms—and a mercantile-financial class focused on the domination of a *space of flows*—the network of nodes and conduits in which production and trade take place. Combined or alternated, these two logics conquered populations, forced them to adopt capitalist relations, and repressed their resistance, establishing the conditions for accumulation; and oversaw the accumulation of wealth, in turn fueling and equipping the military machine that made their economic processes possible.[58] This model parallels earlier theorizations by Mikhail Bakunin[59] and Fredy Perlman.[60]

As we come back to the concept of social war throughout this book, we will see how it provides a necessary framework for understanding the nature of the crisis and realistic responses; how it effectively complements the anticolonial emphasis that is crucial to an honest inspection of ecocide; and also how it explains why the mainstream conversation is constrained in such a way to obscure what is actually going on.

The insurrectionary text, *23 Theses Concerning Revolt*, offers the following affirmation that helps us to untangle the crisis. "Single-issue activism is capitalist alienation transposed on the realm of social struggles."[61] When we understand that enclosure—the privatization, destruction, or sealing off of the commons—is simultaneously a military and economic activity designed to achieve social control as well as to accumulate wealth, we can imagine how enclosure might operate within the movements that seek to save the planet, stop climate change, and so forth.

Single-issue activism on this front means focusing almost exclusively on greenhouse gas emissions, staying "on message" or not "clouding the issue" by intentionally avoiding any deep engagement with Indigenous struggles, colonialism, capitalism, or social war. It means, effectively, enclosing the topic of climate change so that it does not enter into dialogue with a thousand other simultaneous, interrelated conflicts and crises, thus preventing the emergence of a shared field of struggle in which we might put all our grievances, our hurt, and our needs in common, not with a logic of homogeneity or sameness, but precisely with an ecosystemic logic that recognizes the possibility for a mutually beneficial interrelation of differences.

This distinction effectively draws the line between a technocratic, capitalist response to the ecological crisis, and a revolutionary, anticolonial one in which all changes are on the table, everyone is a legitimate actor, and no oppression need remain untouched or unchallenged.

Vital to the conversation around the nature of the ecological crisis is the conversation around how we attempt to fix it. The technocratic pretension that climate is a distinct sphere of ecology which in turn is separate from humanity and human, social problems, is contradicted by one final connection I would like to trace: the connection between the destruction of the earth and our own mental health.

Questions of depression, anxiety, and withdrawal are typically excluded from mainstream discussion of the ecological crisis, but that is a fatal flaw in how society understands the crisis. Over half of child psychologists in the UK reported that their patients experienced anxiety due to ecological problems and the destruction of the environment.[62] Meanwhile, suicide is up among children and teenagers in the US.[63] Likewise, forced migration, which is increasingly being caused by ecological problems, is a major source of emotional suffering.[64] On the other hand, living in relation with the natural world has proven benefits for people's mental health.[65]

I bring this up because I have seen what generalized depression does to social movements. When a cause looks hopeless, people tend towards one or another kind of avoidance. For some people, that means avoiding the question of the ecological crisis altogether. For other people, it means motivated reasoning: leaps of logic that in other contexts would be transparent to them, but that in this case justify false solutions to the ecological crisis. These false solutions tend to constitute a sort of magic wand, a palliative that lets them believe some higher power will come in and fix the problem for us.

To sum up, we cannot look at climate change alone, because the emission of greenhouse gases is directly affected by innumerable other ecological processes, and the ability of the planet and its inhabitants to adapt to the warming, desertification, and sea level rise already under way depends on the health of ecosystems, the presence or absence of specific species, and the impacts of human economic practices, knowledge systems, and legal regimes. Following this thread, every aspect of the ecological crisis is caused by certain practices of human socioeconomic systems, each of which replaced other human socioeconomic practices as part of the global process of colonialism. Some of those suppressed practices constitute effective forms of ecological stewardship that harmonize the needs of the ecosystem and the needs of its human community for health, happiness, and freedom.

HOARD THE PROFIT, SHARE THE BLAME: THE ANTHROPOCENE RECONSIDERED

The changes associated with the ecological crisis are so huge, many geologists favor naming a new epoch beginning with the appearance of anthropogenic or human-caused changes to the planet, whether deforestation, carbon emissions, or nuclear radiation. This is a big deal. The geological epoch that still officially encompasses the present moment, the Holocene, has been going on for 11,650 years and without anthropogenic climate change would probably have lasted for a great deal more time. The epoch before it lasted for two and a half million years, and the one before that lasted for three million years.

The proposal for the name of the new epoch is the Anthropocene, the geological age defined by humans, *anthropos*, even down to the rocks and the chemical processes of the planet. How fair, how accurate, is it to blame these changes on humans?

Though the society that is causing this major ecological crisis is truly global, it only became global through a particular course, and not everyone has the same relationship to it. It might not be "my" society in the same way that it is "your" society. To understand why this is so, we need to take a little jaunt through the history of the ecological crisis and the society that has produced it.

Homo sapiens diverged from another early hominin 300,000 years ago. However, human-related problems capable of destroying life systems on a planetary scale only began to emerge some two hundred years ago. What's more, human social and technological evolution is nothing like a smooth, linear progress from less to more ecocidal behaviors. The society that today reveals itself to be undeniably ecocidal is not some machine that was 300,000 years in the making; the vast majority of human history did not contribute to the present moment, and the idea that human technology is synonymous with ecocide is baseless and, on inspection, racist. It erases the vast majority of human experience, especially that which is external to Europe and its cultural offshoots.

Because a planetary ecological crisis is unprecedented, we can try to find its roots in continental and regional-scale ecological crises. In fact, such crises go back about four thousand years, or just over 1 percent of human history. And because the idea of a unilinear human history is such a pervasive myth, we should also be clear that ecocidal behaviors do not constitute the most recent 1 percent of human history. Rather, ecological crises were extremely rare three and four thousand years ago, happening in just a tiny portion of humankind's overall territory.

What kinds of human societies have caused the collapse of entire ecosystems? As it turns out, there is a clear pattern of major civilizations destroying their soil through deforestation and overexploitation, and then experiencing some kind of political or population collapse. Sing C. Chew documents how, throughout the history of Chinese civilization, economic expansions correlate with increases in catastrophic flooding; both in turn are related to deforestation.[66] Deforestation dramatically decreases the absorption of rainfall, leading to flooding, and economic expansion in an accumulative, statist society means the construction of massive palaces, temples, and navies, the clearing of complex forest gardens for denuded, monocrop fields, and robust fuel consumption not just for human needs but for export economies based on metal, brick, and ceramic production. Deforestation can also spell the disappearance of resources on which a city depends, as when the capital of the Japanese state, Heian, was largely abandoned in the twelfth century CE after the surrounding forests were all cut down.[67]

A multifaceted ecological crisis was perhaps the major cause of the decline of the city-states of Classical Greece. Deforestation, loss of soil fertility, water and air pollution, the local extinction of species like the lion and leopard, localized climate change towards greater aridity and higher temperatures, the subsequent spread of tropical diseases like malaria, all were caused by the state-building and economic accumulation of the Greek city-states. Cities like Athens built commercial empires through colonial outposts from the Black Sea to Iberia; through trade or war and enslavement they made local populations dependent on manufactured goods produced in the center, and in turn had them send back metals, grain, timber, hides, wool, meat, perfumes, and enslaved humans. Athens and the other metropolises were dependent on the colonies for most of their food needs, especially as deforestation in Attica, Macedonia, and Peloponnesus sharply decreased domestic agricultural productivity. They also had to feed an inflated work force of artisans and slaves, such as the 11,000 laborers who toiled in the 87 miles of tunnels of the Laurium silver mines to dig up a metal of no biological utility. All that accumulation attracted the greed of neighboring states and sparked internecine disputes for dominance: Athens practiced genocide against cities that defied it, but their heavy-handedness only provoked more resistance. Attica was definitively deforested to build the warships that would defeat the Persians (449 BCE) and that would sink to the bottom of the Aegean and Ionian seas in the Peloponnesian War (431–404 BCE). In that same century, Athens would have to export its metallurgy industry to be closer to remaining wood sources. Attica lost about half its population between 431 and 313 BCE. The silting of harbors from erosion triggered by deforestation was so extreme that multiple port cities lost their access to the sea: the coastline moved out, and coastal cities like Priene, Myus, and Ephesus became inland cities.[68]

There is also evidence that deforestation contributed to the collapse of the Roman Empire, something the government itself recognized when the Empire started taking measures—too little, too late—to preserve their forests.[69]

All these examples concern ancient states. States are relatively rare in human history, and only in the last few hundred years have they come to dominate the entire globe. But one thing it seems all states do is destroy their environment.[70] States need to be able to rule a subject population and exploit wealth from them. This leads them to impose monocrop field agriculture. Monocrop fields are easy to tax because they all become ripe at the same time, the amount of tax due is easy to calculate on the basis of simple geometry (much of which was developed for exactly this purpose), and the subject population cannot simply run away from state territory—a primary response of people to state authority right up until states became universal—because their food security is tied to fields that can easily be found and destroyed. On the contrary, stateless populations that have to exist in defiance of state authority tend to rely on diverse practices of food autonomy, favoring a mix of hunting, gathering, forest gardens, and root crops that are easy to hide, easy to disperse through a non-flattened, "illegible" stateless landscape, and that can be harvested when most convenient for the harvester.[71]

Incidentally, and in a clear confirmation of the usefulness of the concept of social war, state-resistant food practices encourage people to exist sustainably as part of a robust ecosystem, reject authoritarian forms of organization, and facilitate defensive guerrilla warfare, whereas statist practices of monocrop agriculture make it easier to surveille, extort, and control subject populations, destroy biodiversity, and they eventually lead to ecological crisis.

There are also plenty of examples of states passing laws to protect portions of their environment, though "states will act to prevent environmental degradation only when their economic interests are shown to be directly threatened,"[72] and, I would add, when their social control or "national security" are threatened. This has several implications. Often, because of the nature of tipping points, states tend to take action too late. Furthermore, the purpose of state conservation is never to restore ecological health or even to protect the survival and quality of life of its subject populations, it is to preserve its wealth and power. This is one of the origins of the very concept of conservation: roping off an area of designated natural resources to protect it from human usage, though at the same time to preserve it for other usages the state deems strategic. Conservation, therefore, tends to be an act of enclosure, an attack on the very practices of commoning that actually make it possible for humans to respectfully take part in an ecosystem. The French state provides a good example with the creation of the artificial Landes forest throughout the nineteenth century, draining off extensive wetlands that supported a largely autonomous pastoral population in favor of pine plantations to benefit the timber and turpentine industries.

There are other grave implications to the logic of state-driven conservation. Since the objective is strategic, state conservationism is accompanied by ecological imperialism. The game is about protecting your domestic resources, while destroying or occupying the resources of your neighbor. The British Navy—which was the primary vehicle for Britain's massive colonial empire—ensured a supply of large trees needed for ship construction by expropriating communal forests and excluding peasants from the forest ecosystem they had long relied on, on pain of death. The government even began applying the death penalty to anyone who obscured their face with charcoal, a practice peasants had taken to so as to be able to gather firewood or hunt in anonymity.[73]

Bestseller Jared Diamond has told a story about forest conservation during Japan's Edo period that has become something of a parable for many environmentalists desperate to believe that the *right* government can flick a switch and begin protecting natural resources as a sheer act of will or mere policy divorced from any other considerations.[74] What actually happened is more complex. Statist militarism and warfare were major drivers of deforestation in sixteenth century Japan. After a bloody series of wars to establish centralized control, the Tokugawa Shogunate tried to preserve state power and prevent collapse by encouraging feudal relations between classes. The military/political class had rights of exploitation, but subject to clear limits, and the laboring class had a high degree of economic autonomy (as compared both to imperial slave economies and liberal modernity). The shogunate passed laws preserving the mountainous areas as forest land—in part hunting reserves for the elite and in part village commons—but it was local village councils and not the government that carried out the actual practices that restored the decimated forests.[75] In other words, the state undertook these measures to provide a social release valve and avoid the combined threats of social collapse and economic collapse. They did not have the capacity to carry out their dictates or manage reforestation. It was largely left to common people to do this.

The feudal compromise allowed the Japanese state to survive, and the next chapter in the story illustrates why it is not a good idea to entrust states with power over the environment: because state policy reflects the state's interests, not our own. With the Meiji restoration, state power centralized rapidly and industrialization proceeded in leaps and bounds as the fledgling empire sought to compete with neighboring states and intruding Euro-American powers (militaristic competition being an inherent problem where states are concerned). The Meiji government legislated forest protection, but these protections were on paper only—another recurring theme when it comes to state-led conservationism—and the bountiful forests of the Edo period were quickly cut down for railroads, building construction, and factories.[76] In tandem, the Japanese state began an aggressive drive towards resource imperialism, conquering Taiwan, Korea, Manchuria, and other territories, and setting itself on a path to carry out some of the major genocides of the twentieth century. In the post-war period, Japan was able to replenish its depleted forests around the time its emerging high-tech economy favored the importation of cheap forestry products from poorer neighbors that could not afford the luxury of ecological protections in the new neoliberal economy. The Japanese market has been one of the major drivers of deforestation in the tropical forests of Malaysia and Indonesia.

While it seems that all states are ecocidal, not all societies that damage their environment are states. However, social hierarchies still play a role. The Rapa Nui society (called "Easter Island" by colonizers) experienced a population collapse when their soil failed due to deforestation around 1600. Rapa Nui oral history speaks of a more centralized, hierarchical political system before the decline. Carving and transportation of the *moai*, stone statues that elevated the elite status of clan leaders, required extensive scaffolding and especially rolling logs, and was one cause of deforestation together with extensive agriculture. Deforestation was so extreme that several tree and other species went extinct and the topsoil eroded, leading to severe food shortages. I should point out that several white writers have likely exaggerated the extent of the collapse and spun lurid tales of cannibalism, despite a scarcity of evidence of increased interpersonal violence.[77] It is telling that white writers tend to place all the emphasis on the population decline that occurred before European colonization of Rapa Nui, and not on the collapse caused by violent colonizers. The former resulted in a population decline by a factor of about five, the latter—European colonization—resulted in a population decline by a factor of over twenty.

States tend to lock people into fixed ways of living in demarcated territories, even if those ways of living are unsustainable, in order to hold onto power. Stateless societies are more able to change their ways. Of the three earliest civilizations—societies based on irrigated field agriculture and at least some prevalence of city construction—one of them, the Indus Valley or Harappan civilization, appears to have been stateless, and certainly exhibits far fewer archaeological indicators of statism than the Nile Valley and Mesopotamian civilizations, such as monuments to rulers or supreme deities, wealth inequality, food poverty among the lower classes, and a militarized architecture or other signs of permanent warfare.[78] If they had a ruling class, the rulers left no sign of their passing, in stark distinction to all known ancient states.

Rather than entering into a cycle of imperial expansion and eventual collapse, like the other two, the Harappan civilization was the only one to walk away when ecological conditions no longer favored large field agriculture and dense urban living. In this case, the crisis was probably not one of their own making (also a relevant distinction), but a natural interruption of the monsoon cycle and a decrease in precipitation. In response, Harappan society migrated upland to a territory that still had sufficient natural irrigation, and shifted to a more decentralized, small-scale settlement pattern.[79]

This response is especially relevant when we compare it to Mesopotamia, which existed as part of the same world system as the Indus Valley cities, enmeshed in intense networks of trade. Deforestation linked to economic expansion in Mesopotamia caused increased flooding, the silting up of irrigation channels, and the salinization of fields, making it increasingly difficult to grow crops. In the Indus Valley, rainfall decreased around 2200 BCE, by which time the Harappan cities had already begun a slow decline and after which they changed their settlement patterns more rapidly.[80] This suggests they began to give up on their more exploitative, labor-intensive lifestyle as soon as the ecology became unfavorable, perhaps even starting this transformation before the climatic shift. Yet the decrease in precipitation in Mesopotamia began earlier, with important droughts throughout the fourth millennium BCE and again around 2200 BCE,[81] yet the advanced bureaucratic state took no notice and forced its subjects to carry on with business as usual. The ecological crisis eventually became unavoidable. In the Early Dynastic Period, 3000 to 2350 BCE, crop yields stood at 2,030 liters of grain per hectare. By 1700 BCE, this had dropped to 718 liters per hectare. But rather than read the writing on the wall, elites simply forced lower classes to work harder in squeezing every last drop of blood out of the stony earth, such that cultivators began seeding more intensively and skipping the vital fallow years, leading to further degradation of the soil. The state mandated surplus production, captured entire populations, and put them to work in towns focused on large-scale textile production to keep their economy humming. It was a lost cause: the cities of southern Mesopotamia faded away, and power shifted to the north with the rise of Babylon.[82] But Babylon had learned nothing, and its days were also numbered. Much of the region of what was once lush forest and fertile gardens is still a desert today.

There may have been processes of adaptation and abandonment similar to the Harappan experience in the Mississippian, Maya, and Tiwanaku civilizations of the Americas. Around 1000 CE, a large civilization arose along the Mississippi river based on the cultivation of maize. They built large earthen pyramids that played a role in a more centralized, hierarchical spiritual system. Perhaps in response to the declining quality of life among the incipient lower classes due to state effects like increasing warfare and stratification, deforestation and biodiversity loss, possibly compounded by climatic changes, the civilization was abandoned around 1400–1500 CE and the population diffused to live in a more decentralized manner, selecting which of their society's technologies—from maize cultivation to mound-building—to preserve and readapt on a smaller scale.

The Maya civilization erected huge cities and great monuments across the Yucatan Peninsula and beyond. Millions of Mayan people still live in the same region today, though Western journalists, historians, and educators tend to speak about a "disappearance" or a "lost people." This Indiana Jones-style trope combines the racist erasure of the Other with the tendency of statists to view stateless periods of history as "Dark Ages."[83] Amid all the scholarly articles on the "Classical Maya collapse" of 800–900 CE, it is difficult to find any that take into account oral histories passed on by Mayan peoples today. What occurred was not a disappearance, no one vanished, rather they changed where and how they lived. Most notably, over vast areas they stopped building cities and high-status monuments, and no longer submitted to the political authority of theocratic rulers. Significantly, "regions that were less integrated and hierarchically organized in the Classic period experienced less dramatic collapse."[84]

The causes included a long drought caused largely by deforestation and exacerbated by soil depletion from large-scale agriculture, and increasingly lethal warfare as different monarchies struggled for dominance. Biodiversity loss was also a factor, as deforestation led to the collapse of many animal populations, the disappearance of opportunities for hunting, a less diverse diet with less protein, and thus greater susceptibility to famine.[85]

Some kind of popular rebellion also played a role, whether a massive refusal or an all-out revolution, as people abandoned or toppled the ruling elite and in at least some areas refused to maintain centralizing spiritual, political, and agricultural practices, instead favoring more decentralized, egalitarian, ecological practices, or in the words of one study, "less tolerance for hierarchy and centralization" paired with a "collective ethos" and an ability to organize collective works.[86] This was more an adaptation than a collapse, as they preserved much of their spirituality, arts, and knowledge systems up to and beyond the Spanish invasion. The social mobility that allowed them to survive the havoc wreaked by their governors also allowed some of them to escape Spanish domination for nearly two centuries by moving to the geographies most illegible to European colonization.[87]

The Tiwanaku civilization in the Andes existed from about 550 to 1000 CE, erecting stone cities dominated by great temples. It was a theocratic system in which a diversity of gods and spirits were unified under the central authority of the Sun God, access to whom was controlled by a professionalized caste of priests. Andean oral history speaks of a primarily spiritual rebellion against Tiwanaku centralization and an abandonment of the civilization that left the region stateless for several centuries.[88]

In a contrasting example, social elites may take advantage of ecological crises they cause in order to augment their power. It seems possible that in both Ancient Egypt and Hawai'i before colonization, an ecological disaster created a population of environmental refugees who could be exploited in ways that popular social values previously would not have allowed, granting elites access to a subject population and a dependent labor force necessary for accumulating wealth and power.[89]

We have found one piece of the puzzle. Humans do not destroy the environment, but states do. Stateless societies are by no means perfect, but they are less likely to carry out ecocide and more able to adapt and change their ways. Entrenched hierarchies inside and outside states often encourage ecocidal behaviors and prevent adaptation. Throughout human history, revolution has been a commonsense and effective response to the ecocide conducted by ruling classes. *If we do not revolt in the face of ecocide, the likely result is that states will increase their power and intensify their exploitative practices.*

The regional ecological collapses we have just looked at present a clear precedent to the kind of ecocide we are faced with today, but we must find another cause for two important changes: the extreme quantitative shift from regional disasters to a global disaster; and a qualitative shift from cyclical histories of centralization and rebellion to a linear history of exponential growth in which all territories are locked into an insatiable process of accumulation and ecocide.

Industrialism is an obvious candidate, but I would argue that placing the blame on a society of smokestacks and fossil fuels is to confuse cause with effect. Massive deforestation in North America was already well under way before the widespread deployment of steam power and the advent of industrialism. Early settlers, scientists, and politicians, from Christopher Columbus to Thomas Jefferson, advocated deliberate deforestation to "improve" the climate.[90] Meanwhile, sixteenth-century deforestation in Britain fueled the emigration of poor settlers to the colonies.[91]

Industrialism was the consequence of the exponential expansion of capitalism. The engine for that expansion was colonialism. In order to take off, industrialism needed access to liquid capital for large-scale investments, it needed captive markets, and it needed the reliable availability of cheap resources like sugar, cotton, and animal skins. These were the products of colonialism, from the Spanish silver mines of Potosí that used Indigenous slave labor, to the Spanish, French, and English sugar plantations of the Caribbean and the cotton plantations of the North American mainland (all of which relied on the mass enslavement of Africans), to the French fur-trapping empire farther north, to the cycle of accumulation engineered by the Dutch in Southeast Asia.

The cause of the global ecological crisis is colonialism. It is no coincidence that the political, economic, and cultural institutions that were developed by the most successful northern European colonizers are the ones that are now global. These institutions—from stock exchanges and corporations to universities, land privatization, political parties, and property-based legal systems—were spread throughout a world system that was created by wars of conquest and acts of genocide, with all other ways of life being brutally stamped out. The British, whose empire came to cover 25 percent of the planet's land area, used an effective combination of tactics. They conquered footholds in territories they wished to colonize, decimated local populations in vicious acts of warfare, gave defeated societies the chance to become British allies, sold them European weapons to serve as effective proxies for the capture of slaves and territory, imported enslaved peoples from other areas to jumpstart a plantation economy and destroy any practices of commoning, and further integrated colonized peoples with missionaries and with treaties that offered the hope of survival—adapting to European ways in order to be recognized as human by European overseers—even as treaty-guaranteed rights were progressively whittled away and forgotten. Other colonizers, from major players like the Dutch to relatively minor players like the Swedish, used similar methods.

The kind of economic growth that accompanied colonialism was extractivist and genocidal. From the islands of the Taíno to the Andean highlands, Spanish colonizers forced Indigenous peoples to mine gold and silver, cutting off their hands if they did not meet their quotas.

Belgium applied similar tactics in central Africa. When part of the population of the Maluku Islands refused to continue supplying spices for the Dutch, the Dutch massacred them and imported an enslaved population from other islands, building a new society from the ground up, designed for maximum productivity and discipline. When the Dutch, Portuguese, British, and others could not conquer China outright, they conquered trading posts and began to organize large-scale opium production in South Asia, importing the addictive commodity in order to jumpstart a capitalist economy. Over twelve million people were kidnapped in Africa and enslaved, sent to work in sugar and cotton plantations in the Americas.[92] One to two and a half million died in the terrible conditions of the transatlantic voyage, and many millions more died from the wars encouraged by European slave traders in order to get their captives. Europeans also began a profitable commerce that continues in Africa to this day with murderous results: the international arms trade.

Cotton went to supply the looms of the first industrial proletariat on the planet, many of them child laborers. Marshlands were drained and forests were cut down for fuel, for the building of ships, and for the establishment of plantations. The machine kept accelerating; next it turned to coal as a fuel source, and the destabilization of the planet's climate, already begun thanks to deforestation, was now fully under way. The society responsible for colonialism and for capitalism would throw anything and anyone into its furnaces, whatever the consequences.

Such were the origins of the global economic system we have today: resources extracted by the most ruthless means available, rationalizing the pillaging of living ecosystems and the crushing of human spirits and aspirations so that everyone and everything that does not belong to the owning and governing classes is squeezed for profit or left to rot in surveilled sacrifice zones; natural wealth shipped across the globe to be transformed by machines and workers treated like machines, turned into commodities and sold for a profit that goes not to the ones who did the work but to the ones who were given the privilege of organizing these circuits of exploitation; profits reinvested in similar exploitative ventures, or in fatuous gambling on the likelihood of such ventures to turn a profit, on future prices and currency values, anything that satisfies the fundamental imperative of capital to accumulate more capital, growth *by any means necessary*.

Never before in human history had capitalist logics been hegemonic in human life and the life of the planet, with a new model of the state, effectively fused with capitalist accumulation, conquering the globe and effectively criminalizing and destroying the commons everywhere, forcing the entire human species out of its myriad ecological niches and into a suicidal dependence on a growth-based economy.

Official decolonization after World War II did little to change this dynamic. Because it had to occur according to the colonizers' timeline and criteria, governed by coups, fresh invasions, debt bondage, and structural adjustment, it is more accurate to speak of neocolonialism than of any substantive end to colonization.[93]

The shift to new forms of colonialism only strengthened the implantation of Western institutions. If colonialism created the possibility for a global ecological crisis, the crisis is currently accelerating not through inertia but because the same global system, in updated form, has intensified. More powerful than ever, the stockholders and governors of this system not only hoard the wealth; they are uniquely responsible for the devastation caused by an extractive economy. The wealthiest one percent is responsible for double the CO_2 emissions of the poorer half of the global population.[94]

While some humans are profiting immensely off the destruction of the planet and doing everything they can to prevent meaningful change, other humans are doing everything they can to protect the land and preserve healthy relationships between humans and the rest of the ecosystem. Consider that 80 percent of biodiversity on the planet is to be found on Indigenous territory. Saying that "humans" are responsible for ecological devastation is a continuation of colonial racism, and it is an insult to the peoples who have fought against obliteration to preserve their way of life and their relationship with their territory. It is also an insult to the many people who, despite growing up in a culture totally infused with the values of capitalism, have risked their lives and freedom to defend the land and halt destructive development projects. And it is an insult to the hundreds of millions who are subjected to extreme poverty or absolute precarity by the very same economic order that profits from ecocide, who have to worry about their personal survival and that of their family and community, and do not have the luxury of choosing between different job opportunities and consumer products based on how "ecological" they might be.

As Kathryn Yusoff argues in *A Billion Black Anthropocenes*, the framework of the Anthropocene is racist, and it also serves to obscure our view, to hide the actual system at the heart of the problem.[95] People in the Global South—people dehumanized by Western slavery, colonialism, and racism—finally get included in the category "human," just in time to share the blame for the devastation caused by a social system that has ravaged them far more than they have profited from it.

The ecological crisis is not only the greatest crisis we face, it also encompasses nearly every other crisis and conflict, including the economic crisis that subjects billions of people to precarity or crushing poverty, the crisis of legitimacy that is plaguing governments worldwide, and the technological crisis that juggles problems of totalitarianism, surveillance, and mass unemployment.

2
Foxes Building Henhouses

GOVERNMENT PROMISES AND MARKET SOLUTIONS: THE PROFITABLE FAILINGS OF PARIS, THE NGOS, AND CLIMATE CAPITALISM

Government responses to the ecological crisis are characterized by the compartmentalized, single-issue approach that we have just seen is inadequate. Climate reductionism—reducing the ecological crisis to a simple (and technocratic) question of atmospheric carbon—makes perfect sense to governments, because if every aspect of the problem is governed by a different treaty or legal framework, they can minimize their efforts, focusing only on the issues prioritized by the media and wealthier constituents. Though warming and climate migration are part of the same issue, governments can make grandiose promises about reducing emissions even as they further militarize the borders responsible for so many deaths.

Yet government responses have also been a catastrophic failing for the issues at the center of their reductionist approach. To get a sense for this, we first need to understand why it is such an unmitigated disaster that the United Nations Framework Convention on Climate Change adopted the much-lauded Paris Agreement in 2016, and not, say, in 1978.

Climate change is not a new idea. We have already seen how colonizers in the Americas, from the fifteenth to the nineteenth centuries, advocated deforestation in order to intentionally cause a change to regional climates. As for global climate change, that has been a concept for more than a century. In 1876, anarchist geographer Peter Kropotkin made a connection between industrial pollution and the melting of Siberian glaciers.[1] Twenty years later, the Swedish scientist Svante Arrhenius quantified the role carbon dioxide played in keeping the planet warm and suggested that the industrial burning of coal might produce a "noticeable increase."[2] In 1938, Guy Stewart Callendar, a British engineer, published evidence that CO_2 levels and temperatures had been rising over the prior fifty years. In 1959, scientists projected that atmospheric carbon would increase by 25 percent by the end of the century, causing major changes to the climate.

President Johnson's Science Advisory Committee warned in 1965 that the burning of fossil fuels could create a runaway greenhouse effect, and in 1968 the Stanford Research Institute predicted the melting of the ice caps, the rise of sea levels, and a substantial increase in temperature that could likely affect global climate by the year 2000. Incidentally, Stanford carried out their report for the American Petroleum Institute. In 1967 and 1968, Exxon and Shell Oil, respectively, filed away internal reports that their activities were causing the air pollution that today is responsible for millions of yearly deaths.[3]

In 1969, NATO began studying climate change as a possible security threat. At this point, most climate scientists agreed on global warming, though there was not yet a consensus. A survey of scientific studies published between 1965 and 1979 found that 44 predicted global warming while only 7 predicted cooling.

Droughts in Africa, Ukraine, and India in the early '70s caused major food shortages and spread concerns about the possible effects of climate change. In 1978, the World Meteorological Organization announced their agreement with the greenhouse theory, and in the following years, further measurements and discoveries gave an increasingly convincing and complex picture of anthropogenic climate change. In 1988, hundreds of scientists gathered in Toronto and concluded that pollution caused by humans was already having "harmful consequences" in many parts of the world. They recommended a 20 percent reduction in emissions by 2005. That same year, the WMO and the UN formed the IPCC.

In other words, we already had strong evidence in the 1950s that climate change caused by fossil fuels was a serious danger threatening the planet, and comprehensive studies in the 1960s showed that this view was *probably* correct. By the '70s, we can speak of a scientific consensus on climate change that was consolidated throughout the '80s.[4]

Nonetheless, it was not until 2001 that the consensus was finally announced to the world in a very public way, as 34 different national science academies and two important international scientific councils formally declared in favor of this view, as did the IPCC in its third assessment report. From the moment when a strong majority of climate scientists were predicting dangerous global warming, including those funded by the petroleum industry, 20 years went by before the UN devoted significant attention to the problem, 33 years went by before the international scientific community announced their consensus in a highly public way, and 37 years went by before the first international treaty purportedly aiming to reduce greenhouse gas emissions went into effect. As for governmental treaties that have actually led to greenhouse gas reductions, we are still waiting, more than sixty years later.

The lag time—half a century between a strong scientific majority and some semblance of coordinated government action—shows how dependent our institutions are on the extractivist industries responsible for the ecological crisis.

The media present an extreme case: in the 1970s, mainstream news articles about climate change abounded, but they were all warning people about global cooling, claiming that meteorologists couldn't keep up with all the new evidence. In the '80s and '90s, corporate media covered the topic with a blanket of silence as they shifted to a more subtle strategy of contention, treating "both sides" of the debate as equally valid, giving huge platforms to scientists in the pocket of the oil industry who claimed that global warming could be beneficial, or that such changes were "natural." Eventually, they shifted to delaying tactics, and the media and the fossil fuel lobby boosted the kind of naïve environmentalism that put all the focus on individual consumer choices.[5]

Simultaneously, conservative media outlets accelerated their campaign of climate denial and deliberate misinformation, spreading ludicrous conspiracy theories about environmental activists somehow profiting off of their opposition to the multi-billion-dollar petroleum industry. The hypocrisy is staggering, considering that just five major oil and gas companies spend $200 million every year "on lobbying designed to control, delay or block binding climate-motivated policy."[6]

Capitalism in general has plenty of mechanisms to indirectly control a nominally free media and guide policy debates in democratic governments.[7]

Up until the end of 2006, 52 major news sources that were polled across the world only mentioned climate change a couple times a month. Media interest spiked in 2010, but between 2011 and 2015 climate change fell out of the news again, almost as low as the pre-2006 levels.[8]

In two weeks of hurricane coverage in the US in 2017, eight major TV networks mentioned climate change only 5 percent of the time. In other words, people were losing their homes and hundreds were dying, but on the whole the media refused to educate people to the fact that climate change has been making hurricanes more frequent and more deadly.

"ABC and NBC both completely failed to bring up climate change during their news coverage of Harvey, a storm that caused the heaviest rainfall ever recorded in the continental US."[9]

Longer form media also tends towards misinformation. In 2020, as I was beginning research for this book, not a single one of the top 10 titles on "climate change" recommended by Amazon was an evidence-based work about the dangers and possible solutions. On the contrary, what came up were industry-promoted bestsellers about how climate change was not such a big danger, or that protecting the economy was more important. The closest thing to a sensible perspective was a sort of self-help narrative on dealing with the anxiety we may feel about our dying planet.

Inaction and prevarication on the part of government, media, and other private companies seem to be a case of the wealthy and powerful not taking the problem seriously, but that explanation does not fit the available evidence. Energy companies are both informed and sober as they study oil futures and decide whether to invest in fossil fuel infrastructure or alternative energies. Insurance companies are deadly serious when they calculate risks and losses from warming-aggravated disasters. And it was more than 60 years ago that Western governments first began studying climate change as a security threat. It only seems like leading institutions have not taken the crisis seriously because their response has been so thoroughly sociopathic. They have accelerated fossil fuel extraction, racing to achieve as much economic growth as possible before public opinion turns against them, and then insisting they sit at the head of the table when environmental protection finally becomes a serious policy issue, though the proposals they spin out are all designed to achieve even more economic growth, this time under a thin veneer of toxic green paint. At no point has it occurred to corporate and governmental leaders to empathize with the millions of humans who are dying, suffering needless diseases, hunger, and thirst, or to contemplate the tragedy of species and habitats that are disappearing forever.

Within this historical framing, it also helps to see that the only thing that separates corporate bureaucracies from public bureaucracies is a revolving door. Whether key companies are state-owned, as in Norway, Indonesia, or China, or whether they are just subsidized by the state and entrusted with authoring regulations themselves, as in the US, the ruling class and the owning class are thoroughly enmeshed.

One more piece is missing to understand the highly effective disaster constituted by the official solutions to the ecological crisis: the opposition. The NGOs that dominate the environmental movement are no less integrated into the capitalist industries destroying the planet than the most corrupt Republican or Tory official.

At an increasing pace throughout the '90s, environmental NGOs shifted to a strategy of partnership with big business, increasingly accepting corporate donations and leasing out their logos to grace the products of Coca-Cola, Clorox, British Petroleum, Shell Oil, Unilever, McDonald's, Boeing, Monsanto, Starbucks, Walmart, and Ikea to convince consumers to keep shopping. Each of these companies have received the approval of, and generously donated to, NGOs like World Wildlife Fund, Sierra Club, the Nature Conservancy, the Environmental Defense Fund, and Conservation International. Even Greenpeace, one of the few that does not accept corporate donations, lends extremely destructive corporations like Unilever its good reputation through partnerships. The relationship between NGOs and corporations has become so tight, many share board members, like BP and the WWF. The Sierra Club accepted over $27 million in donations from the gas industry between 2007 and 2010, while the Nature Conservancy has accepted millions for the ludicrous project of helping BP carry out "sustainable" oil extraction.[10]

There are small NGOs that are effective at providing resources to those with immediate needs of survival; those that operate boats rescuing migrants from drowning in the Mediterranean come to mind. But to avoid the age-old dynamic that sees little fish turn into big fish, the entire context needs to be radically shifted. Grassroots movements are resourceful, but there is no doubt that we need more resources on the ground. However, it is never worth it if we have to give up our autonomy and our revolutionary horizons to achieve those resources. And selling out is a major industry. Social critics have already drawn attention to the "non-profit industrial complex."[11] The higher the stakes are with climate change and the ecological crisis, the greater the institutional pressure to sell out, and the greater the reward.

Greenwashing pays off. The CEO of the Wildlife Conservation Society was paid $1,636,145 in 2019 while the CEO of the World Wildlife Fund was paid $1,455,444, putting them high on the list of highest paid charity CEOs (especially when one considers how many of the other charities on the list are connected to the profitable medicine, research, and real estate industries).[12] These salaries, already crass profiteering off the dire situation of life on this planet, become even more shocking when we see how they have evolved. Ten years earlier, in 2009, the CEO of the US branch of the WWF was paid $455,147.[13] What caused his salary to triple? Did the big NGOs suddenly get more effective at forcing governments and business partners to protect the environment? On the contrary, the last two decades have been a solid streak of missed targets and broken promises from biodiversity to carbon emissions and more. What are these unjustifiable pay raises reminiscent of? The financial bosses who also saw their salaries skyrocket even as the monetary system they supposedly safeguard tumbled into recession, leaving the rest of us battling unemployment, evictions, and pay cuts. There should be no mistake: the NGO elite are a part of the mercenary apparatus that is selling the planet piece by piece.

Now that the stage is set, we can watch the performance. The Paris Agreement, and the Kyoto Protocol before it, seem to be little more than publicity stunts. Even as the 2005 Kyoto Protocol was touted as a success, global emissions rose by 32 percent from 1990 to 2010.[14]

The Paris Agreement, with 189 signatories since it was launched in 2015, sets a goal of limiting climate change to 2°C above pre-industrial levels. It is hard to know whether to criticize Paris for its illusory goals or its insubstantial methods. The Agreement does not include firm commitments and enforcement mechanisms, an omission reflected in the unfortunate fact that not a single major industrialized member has met its emissions targets.[15] Yet even if all these governments started keeping their promises, and that's a big if, 2° of change comes with a fairly high possibility of triggering several interconnected tipping points and moving to a "hothouse" planet. According to a 2018 study, "We note that the Earth has never in its history had a quasi-stable state that is around 2°C warmer than the pre-industrial and suggest that there is substantial risk that the system, itself, will 'want' to continue warming because of all of these other processes—even if we stop emissions."[16] Specifically, the Agreement seeks to stabilize atmospheric CO_2 at 450ppm. Never mind that this comes with a 26 to 78 percent risk of pushing well past 2°C of warming.[17] Acceptable odds, for world leaders. Many economists are happy, though: their cost-benefit analyses suggest that a 3°C temperature rise would be ideal.[18] We can only imagine the low monetary value they place on the lives of the tens of millions of people who are dying, and all the species that are going extinct.

Has Paris made it clear that a new, green economy is on the way? The world's banks don't think so. Since 2015, the 60 largest banks have financed fossil fuel companies to the tune of $3.8 trillion, and the loans are only increasing in quantity.[19] The planet's governments are on the same page. Collectively, they provide $500 billion in subsidies to the fossil fuel industry *every year*.[20] These are the governments we are entrusting with the solution to the problem.

It is no wonder that a Shell Oil executive was caught boasting about how they helped write the Paris Agreement, particularly the sections that allow for carbon trading.[21] As Laura Terzani writes, "net zero emissions is not zero emissions."[22] In practice, net zero usually means wealthy countries passing the responsibility for their emissions on to poorer countries through creative accounting. When we pare down the ecological crisis to the demands of climate reductionism, talking about capitalism is dismissed as "off message." Should it be any surprise, then, that climate policy has just become a new scam allowing the wealthy to make even more money?

It's worse than mere profiteering, as a social war analysis would lead us to suspect. Carbon trading creates a market for major land grabs across the Global South, a final nail in the coffin of the world's peasants, and a continuation of the colonial dynamics that export the costs of the ecological crisis to poorer countries. It is also a key part of "payment for ecosystem services," a way of "quantifying and constructing the natural environment as a 'service provider' commensurable with financial markets," which in turn constitutes a pillar of "property-based counterinsurgency," the pernicious and often brutal repression of movements to defend the land.[23]

Placing our trust in Paris is an extremely precarious proposition. As democracy unravels, fewer and fewer governments are following a consistent course in policy. The worse the crisis gets, the better the electoral calculus of right-wing populism and economic nationalism that can lead to 180-degree shifts in policy. This is a major strategic liability for the mainstream climate movement, as seen in the US and Canada withdrawing from their climate commitments, Modi in India encouraging coal mining and canceling the ban on single-use plastics, or Bolsonaro in Brazil pillaging the Amazon for increased political and economic capital. The more power they have, the less governments need to listen to us.

Even governments that maintain their Paris commitments are making a cavalier calculation, hoping to "ease" the transition and meet their emissions reductions at the last possible moment. But what happens when you have disasters like the 2019 Australian wildfires that released a quantity of carbon dioxide equal to half the country's annual emissions, or the simultaneous massive die offs in Australian mangroves and Tasmanian sea grass forests, each of them more important carbon sinks than the dry land forests?[24] Such catastrophes will become increasingly common, and they make a mockery of the pretensions of governments to be in control. Emissions reduction targets are calculated to let capitalist economies squeak by, making the minimum change necessary at the last moment possible. They represent a climate brinksmanship that will ineluctably send us tumbling over the edge.

On a technological as well as a policy level, the instruments for reducing global warming are a part of the problem, rather than the solution. Because emissions reductions are designed to enable the continuous economic growth on which capitalism is predicated, "going green" means, more than anything else, a shift from fossil fuel energy to renewable energy on a massive, industrialized scale. But renewable energies on an industrial scale are also extremely destructive.

Hydroelectric power is currently the foremost renewable energy source. Hydroelectric dams are extremely destructive to build. They require a huge amount of concrete, production of which is one of the main greenhouse gas emitters, they cause the loss of huge amounts of forest and farmland, they kill off riverine species such as salmon, they disrupt natural flooding cycles necessary to many ecosystems, and their reservoirs emit large quantities of methane.

Photovoltaic cells capturing solar energy—solar panels—frequently use toxic heavy metals such as cadmium, gallium, and lead in their construction. On a localized scale, houses in most climates could be heated, cooled, and provided with warm water through efficient design capturing solar energy. And if people were willing (and able) to alter their daily habits and limit their electricity usage, photovoltaics free of heavy metals could play a role in this. Many of the problems arise when we take the current energy economy and assume we can keep on operating in the same way, but with different inputs. To meet current energy usage, huge swathes of land would need to be appropriated, denuded, and fenced off for solar farms. And since photovoltaics are only productive during sunlight hours, we would have to construct an enormous infrastructure for energy storage and global high-voltage direct current power lines, meaning more toxic mining, more land grabbing, more energy expenditures, and more cancer and other illnesses. Solar panels also present a waste problem, as the typical panel only lasts 25 years. At current growth rates, by 2050 we could have 78 million metric tons of junk panels, and there is currently no good way to recycle them at an industrial scale.[25] Andrea Brock and other researchers argue that, far from a "paradigmatic break," solar energy "is merely the latest iteration of an industrial growth model" characterized by "undemocratic and unsustainable industrial processes, the concentration of corporate power and profits, and externalized waste and pollution."[26]

Geothermal energy, aside from having a low efficiency, requires extensive drilling that can contaminate groundwater. The technology uses pentane, a highly toxic, flammable liquid. Geothermal plants release small amounts of methane and the toxic gas hydrogen sulfide, which causes acid rain. In accidents, geothermal plants can release large amounts, such as occurred at a single Hawai'i geothermal plant during a drilling blowout in 1991, in 2013 after an equipment malfunction, and later in 2018 when the plant was damaged by a volcanic eruption.[27]

Nuclear power is not a renewable energy, given that it expends its fuel source, uranium, though there is a huge lobby seeking to promote it. Countries began adopting nuclear energy in the first place not for its energy benefits, but because it advanced their nuclear weapons programs. Nuclear plants come with a constant risk of meltdown, releasing large amounts of deadly radiation into the atmosphere and potentially making the territory uninhabitable for millennia. In the West, pop cultural representations of the Chernobyl meltdown ascribe the disaster to Soviet incompetence, but meltdowns and near meltdowns at Fukushima, Japan, Three Mile Island, United States, and Loir-et-Cher, France, shine a light on Cold War propaganda and show that no regime is immune to disaster. In fact, over one hundred nuclear accidents have occurred since 1952, the largest share of them in the US. But the daily, effective operation of a nuclear power plant may be even worse than a meltdown. In 2011, 75 percent of US nuclear power sites were found to be leaking radioactive tritium.[28] Depleted plutonium rods have a half-life of 24,000 years, which, for reference, is far longer than agriculture or wheels have existed, more than 40 times longer than the longest lasting state survived, and roughly 500 to 1000 times longer than your typical nuclear storage site goes without experiencing a major leak. Nuclear proponents argue that the rods constitute a small volume of toxic material compared to mine tailing from coal production, for example. They tend to leave out the millions of tons of radioactive uranium mine tailings (11 million tons from a single site in Utah) and the 1.2 million metric tons of depleted uranium produced by uranium enrichment.[29] This radioactive byproduct has a half-life of 4,400 million years (or, roughly the current age of the Earth). Inexcusably, those who developed nuclear technology invented no way to safely store all that waste for the amount of time it will pose a lethal danger to all life, and no such storage technology is even on the horizon. Many nuclear waste storage facilities have been found to leak radioactive compounds into the environment.

Wind power is the second most widely used source of renewable energy. To be able to fully appreciate the kind of damage wreaked by green energy on an industrial scale, we can turn to research detailing the consequences of industrial wind farms in Oaxaca, Mexico. La Ventosa was one of the first towns in Oaxaca to be impacted by wind turbine construction, and is now more than 80 percent surrounded by giant turbines.[30] Turbines were largely constructed on farmland bought up from farmers facing dwindling returns in agriculture. In other words, a precondition for the land grab was forcing local residents from a subsistence economy, in which communities fed themselves and preserved Indigenous practices of commoning, into a global monetary economy that systematically drives farmers out of business.

The combined impact of "economic liberalization" like NAFTA and "climate change legislation [that] promoted foreign direct investment (FDI) in Mexico" led to a boom in wind turbine development.[31] Additionally, the system of *caciques* and *coyotes* was vital to organizing land acquisition on the ground, with all the attendant corruption, fraud, and threats of violence.[32] Overall, it was an involuntary process, with the state building high-tension power lines and other infrastructure in residential areas without even consulting residents: "the people are never fully informed, unless they find out and if need be, stop it for themselves. Otherwise development happens."[33]

Turbine construction damages soil, depletes groundwater, causes erosion and flooding, contributes to deforestation, and the turbines themselves kill off both terrestrial animals and birds, and frequently leak oil, contaminating the soil and drinking water.[34] Local farmers have reported their cattle dying off or no longer reproducing, and residents who live closest to the turbines report "headaches, dizziness, digestive problems, vomiting, nose bleeds, exhaustion and insomnia" possibly due to the noise pollution and vibrations caused by the turbines, as well as fumes from the leaking oil.[35]

Most jobs connected to the wind turbines disappeared after construction was finished, and the best jobs were reserved for foreign experts. Similar to the "man camps" that spring up around fossil fuel extraction sites in North America, leading to a sharp rise in violence against Indigenous women, foreign experts working for the wind farms also embodied predatory practices towards local women, frequently marrying into local families to acquire land titles, shifting those titles to the wind industry, and then moving back to their "real" families in Spain or the United States, leaving abandoned women, children, and broken communities in their wake.[36]

Energy produced by the wind farms in Oaxaca goes to major companies like Walmart, cement producer Cemex, and major food processing and mineral extraction companies, thus allowing some of the most destructive industries and consumer lifestyles to appropriate the language of sustainability and paint themselves green. Their use of wind energy also allows these industries "new possibilities of receiving 'climate' and 'clean' technology funds and loans." In fact, the "La Mata and La Ventosa Wind Park became the World Bank's leading Clean Technology Fund (CTF) project in Mexico."[37]

Energy production is also being linked into "an energy export-oriented model that seeks to power industrial zones, private industry and other countries," providing the electricity for the "*maquiladora* corridor" envisioned as part of the much-criticized Plan Puebla-Panama that favors the US consumer economy through environmental degradation and the exploitation of workers in sweatshops throughout Central America.[38]

Wind energy development in Oaxaca has further integrated locals into the global economy, but the results have been an influx of drugs, a much higher cost of living, rents skyrocketing by up to a factor of ten, and, ironically, more expensive electricity.[39] Most residents reported no social benefit from the development, and bitterly recalled the promises made by major corporations like Iberdrola that resulted in little more than photo-ops.[40]

The kind of development brought by the wind farms has also gone hand in hand with counterinsurgency interventions, for example the "participatory mapping" of Indigenous territory and resources in Oaxaca, organized and funded by the US military with the aid of geographers from the University of Kansas. The mapping was officially carried out with a green capitalist logic, but its primary intention was to "gather intelligence on emerging and asymmetric threats to the United States for the purpose of preparing for conflict."[41] Ironically, one of these asymmetric threats blossomed and forced the cancellation of the mapping project, when the people of Oaxaca rose up in 2006 and took control of most of the state, organizing in a horizontal way largely informed by Indigenous traditions.

Similar conflicts and ecological destruction connected to industrial wind farms have been documented in Scotland, Wales, Sweden, Spain, Greece, India, Kenya, Brazil, Chile, and the United States.[42]

Green energy is often used to give extremely destructive corporations a better image, and also win them access to the subsidies that are the main weapon of climate change legislation from Paris Agreement targets to the Green New Deal. Renewable energies are often directly linked to fossil fuels. Germany is destroying the ancient Hambacher and Dannenröder forests in order to mine lignite—the dirtiest form of coal—in some of the deepest open pit mines in the world. Yet they tout their use of green energy to run pumps and other infrastructure at the coal mines, and they have recently announced they will turn one old coal mine into a giant hydropower battery for storing excess solar and wind energy.[43]

Justin Trudeau, the progressive prime minister of Canada, entered office with promises to respect treaty rights and improve relations with Indigenous peoples, but the central push of his administration has been to deforest and eviscerate the center of the country with highly contaminating tar sands extraction and enforce the construction of oil pipelines through unceded Indigenous territory. Trudeau tried to improve his image by promising that the Trans Mountain pipeline expansion, slated to pump 890,000 barrels of oil a day through highly sensitive ecologies, would come with tax money that would be dedicated to green energies.[44]

Green energy is also used as a justification for unregulated, ruinous mining on the sea floor, with the argument that the destruction of the very foundation of the marine ecosystem will produce the minerals necessary for batteries and other renewable energy infrastructure.[45] Meanwhile renewables are being used to justify the privatization of energy in South Africa.[46]

The green economy is fully integrated with the regular economy in other ways as well. Tech companies from Microsoft to Tesla like to style themselves as leaders in the climate crisis, the ones who will come up with the innovative solutions, a completely new kind of capitalist when compared with the dinosaurs of Exxon or Ford. However, there is no such thing as clean money. Big data depends on fossil fuel companies for a huge chunk of their profit. Cloud computing (led by Amazon, Microsoft, and Google) depends on billions of dollars in revenue from the major oil companies, which in turn depend on AI in order to increase the output of their wells, thus fueling more global warming.[47] And all such companies rely on the capital provided by investors, who make that capital doing whatever turns a profit, no matter how destructive that activity is. Capitalism is by definition an amoral system. That is why, for as long as capitalism has existed, no economic practice, no matter how repugnant—from slavery to the manufacture of weapons of mass destruction to the distribution of addictive substances—has ever been effectively prohibited.

One key to maintaining a consumer economy and the lifestyles taken for granted in wealthy countries is a switch to electric cars. Media and manufacturers conspire to label these as "zero-emissions" vehicles, which is a convenient inaccuracy. Electric vehicles simply switch the point of emissions to the power grid. If that grid runs on coal, electric vehicles also run on coal. And the drive to produce hundreds of millions of electric cars over the next few decades, while a major boon for the auto industry, is a huge strain on the environment, constituting an immense draw on energy needs and more greenhouse gas emissions.

Then there is the question of the lithium, nickel, manganese, and cobalt mining necessary for the massive batteries. Nickel mining, carried out primarily in New Caledonia, Australia, Canada, Indonesia, Russia, and the Philippines, creates mountains of toxic slag. Refinement pumps millions of tons of poisonous sulfur dioxide into the atmosphere.[48] Manganese is mostly produced in open pit mines in South Africa. Exposure during mining and processing can lead to neurological disorders. Cobalt mining, mostly carried out in the Democratic Republic of Congo, is responsible for widespread child labor (with more than 40,000 forced to work in the mines), workers' deaths in hazardous conditions, genocidal acts against Indigenous inhabitants, health problems among neighboring communities that range from birth defects and breathing difficulties to death, deforestation, and habitat loss for threatened species.[49]

Meanwhile, the growing demand for lithium was connected to a far-right coup that temporarily ousted the government in Bolivia, with a toll of 33 people killed and hundreds injured. Bolivia has the world's largest lithium reserves, largely in protected areas. In response to criticism about a possible role, Elon Musk went on Twitter to blithely assert, "We will coup whoever we want."[50] Ousted President Evo Morales was industrializing lithium production and had already signed deals with French, Japanese, and Korean companies to devastate a part of the Uyuni salt flats; however, under Morales' plan, Bolivia would have retained a higher share of the profits.[51]

Lithium mining is experiencing a boom in northern Portugal, where open pit excavations are used instead of harvesting brine. The result is deforestation, habitat destruction, noise and chemical pollution, and water depletion, not to mention gouging out the mountainsides. But because the lithium is going to batteries, EU environmental regulators are giving the mining their full support.[52]

Renewable energy and deceptive solutions like electric cars have a deleterious effect on the environment and on human health. They are being promoted in large part to placate the public with the message that those in charge are carrying out big changes, that the solution is already here, and as a means to funnel huge amounts of public funds to major corporations in a modern-day Marshall Plan to prop up a chronically sick economy. The placebo effect of green energy is evident in media reporting. After noting that 2017 broke all previous records for carbon emissions, a typical article calmed readers with some good news. "There have been some promising trends, like an increase in renewable energy, with jobs in that sector rising by 5.7% between 2016 and 2017, according to the report."[53] In other words: there is good news and bad news. The bad news is, our solution isn't working. The good news is, more and more people are using it!

The cognitive dissonance involved in these formulations is astounding.

Aside from industrial-scale renewable energy, many experts talk about the need to create carbon sinks to pull carbon dioxide out of the atmosphere as quickly as possible. A major study expanding on the IPCC's action proposals demonstrated that reforestation was the most important course of action to stop warming,[54] in other words, planting billions of trees over the next few years.

The problem is, when governments try to mandate such sweeping changes from above, the most marginalized elements in society, and also non-human life that is excluded from personhood, get crushed. Top-down reforestation efforts frequently result in land grabs that dispossess Indigenous peoples and agricultural communities and frequently benefit logging, tourism, and other major industries. They are also incapable of spreading actual forests, which are complex multigenerational habitats uniquely adapted to the local territory. So-called reforestation from above is designed by people who don't even understand the difference between a monocrop tree plantation and a thriving ecosystem with trees, nor do they have a financial or spiritual interest in learning that difference.

Chile provides a key example of a neocolonial country with an extractivist economy: its main sectors are copper mining and logging. On paper, Chile is listed as a heavily forested country, and it uses these supposed forests to economic gain, in the form of carbon credits. In fact, the country has very few actual forests left. Much of the loss can be traced to when the brutal Pinochet dictatorship transferred huge swathes of Mapuche lands to the logging industry, which destroyed most of the remaining native forests and replaced them with monocrop plantations of non-native eucalyptus and pine. The plantations are killing the soil, lowering the water table, and depriving the Mapuche of the species they traditionally rely on for their food, ceremonies, and medicine.

The environmentalist pretensions of logging companies and government ministers is contradicted by a key activity in the Mapuche struggle: chopping down trees in forestry plantations in order to replant native trees or to sow crops and achieve food sovereignty.

Much Indigenous territory is also threatened by eco-tourism, another government-corporate scheme for domesticating and exploiting the land that masquerades as a solution to the ecological crisis. Selling a destination as pristine, in the Western imaginary, requires expelling the local inhabitants or converting them into guides, cleaning staff, and other invisibilized sidekicks to the conscientious tourist. The low quality of jobs in the tourism sector and the need to attract wealthy customers from the Global North reveal this as just another iteration of colonialism, and one fully integrated with the destructive airline and cruise industries. Economic recessions show how unsustainable the approach is: when we treat nature as a monetized resource, the shift from cameras back to chainsaws actually occurs quite quickly when the economy demands it.

Even when the chainsaws can be kept at bay, the practice of conservation has generally meant roping off huge areas and evicting the human inhabitants. This is because the world's dominant institutions continue to hold an aporophobic, aristocratic view of nature that excludes humans. Since, for the wealthy, food and survival appear automatically, their god-given entitlement, nature is something external, distant. It can be set aside for specifically delineated, upper-class enjoyments, whether hunting, sightseeing, or scientific research. They certainly don't contemplate humans as a necessary part of the ecosystem. It should be no surprise, then, that paramilitary park rangers paid for by major environmental NGOs like the WWF have been accused of genocidal practices and acts of violence against Indigenous people in the Congo, Cameroon, Botswana, and elsewhere.[55] This includes the WWF's expropriation of Baka lands in central Africa, which are then commercialized with concessions to logging companies and trophy hunters, paired with violent attacks by paramilitary "park rangers." There are also cases of ex-soldiers from the US returning from neocolonial duties in Afghanistan and being recruited for the *good cause* of shooting alleged poachers in African forests; the prohibition of hunting by San peoples in the Kalahari Desert with the support of the British government; the execution of poor villagers gathering firewood in the Zambezi National Park; and the planting of high-powered weapons or elephant tusks next to executed Indigenous hunters in order to portray them as highly paid poachers.[56] And all this in a context in which lucrative trophy hunting by wealthy white people from Eric and Donald Trump Jr. to the King of Spain is a part of the financial calculations of conservationism.

Nor are such practices exclusive to Africa. Informants near Kaziranga National Park in India are paid $1,000 if they give a tip that leads to the killing of a poacher, leading to an industry that preys on those who continue a traditional relationship with the forests.[57] The semi-nomadic Gujjar of Kashmir follow their herds, staying in different homes in the summer and the winter. For countless generations, they have been stewards of the region's forests. Yet it was the wildlife protection department that began demolishing their homes.[58]

And in Oaxaca, Mexico, at the site of the previously mentioned industrial wind farms, paramilitaries working for the energy companies often work with the police to attack or detain so-called poachers in the areas deforested by the turbines.[59]

A UN Convention on Biodiversity plan to place 30 percent of the world's land area in conservation was the target of an open letter that took issue with how the plan "continues the marginalisation of rural people who will be most affected by its measures [and] ignores decades of research and experience on the social impacts of conservation."[60] Even conservation programs designed to be feminist and decolonial, in opposition to the dominant trend of "green militarisation", tend to be severely limited in what they can achieve, because, as documented by Laura Terzani in the case of an all-female anti-poaching unit in South Africa, "a fundamental change cannot be brought about and operate within existing unequal and unjust structures".[61]

Because all accumulation in a capitalist system accelerates further accumulation, the roll-out of renewable energies is being accompanied by an increase in fossil fuels. A similar thing happened in the '80s, when energy efficiency increased, encouraged in no small part by the consumerist environmental movement of the day, and due to capitalist price dynamics total energy usage also increased. In other words, technological improvements touted as being good for the environment often have the opposite effect. Fossil fuel infrastructure constitutes a huge amount of fixed capital that fossil fuel companies are not going to walk away from, and they have a fair amount of leverage over production efficiency and price—spurred in part by green technologies like carbon capture—to make sure the economy keeps buying and burning.

With carbon emissions not slowing down in time to avoid surpassing 2°C of warming, a large number of climate experts and entrepreneurial capitalists are turning to questions of geoengineering and carbon removal. In *After Geoengineering*, Holly Jean Buck provides a useful review of many of the methods being developed and the difficulties and limitations they entail. Since carbon capture is being sold as a way to *buy more time*, she poses the question, time for whom or for what? Are we buying more time for endangered species, threatened habitats, and vulnerable human populations? Or are we buying more time for governments and companies to continue enriching themselves and to slowly wean themselves off of fossil fuels when it is convenient to their profit margins and geopolitical strategies? As she makes clear, "the polluters are first in line to benefit from carbon capture."[62]

Currently, the predominant strategy for carbon capture is through "natural" methods, namely reforestation, which comes with its own problems. But other strategies, many of them classified as geoengineering, are increasingly gaining attention, and sometimes, funding. Examples include injecting carbon pulled out of the atmosphere into cement or into bedrock, where it mineralizes; liquefying carbon and storing it in old gas and oil wells; constantly seeding the upper atmosphere with aerosols that will reflect sunlight; dusting agricultural soils with ground up minerals that increase carbon retention; or running huge offshore plantations for genetically modified algae to be used as a fuel source. The main real usage of any of these technologies, at the moment, is by the fossil fuel industry injecting captured carbon into its wells to displace oil or gas and increase output, known as enhanced oil recovery.

I agree that we need to consider all of these proposals since it is a matter of stopping mass starvation and saving species from extinction, but only if we can do so fully aware of their true dimensions.

First off, we need to consider the *cane toad phenomenon*, whereby interventionist technologies introduced to solve problems created by earlier interventions create even bigger problems. As Western colonial civilization fumblingly developed an interventionist, mechanical understanding of ecology, in this case in Australia, they released the giant cane toad as a predator to control runaway populations of rabbits, an earlier colonial introduction, but the cane toads ended up eating everything, decimating many native populations.

The truth is, at the intersection between knowledge and business that constitutes almost all of today's scientific practice, there is a solid tendency to pollute first, and look into the health problems later, from plastics to PFOA to phthalates. If you dangle a dollar bill over a cliff, today's science-driven industries will jump and assume they can use their earnings to buy a parachute before they hit the bottom.

A pertinent example of this phenomenon is how carbon capture at scale will require a massive infrastructure including a huge network of pipelines for transporting carbon from its capture site to its storage site. The fossil fuel industry are the only players in the current context with the ability to build—and profit from—this infrastructure.

And the reality is, all pipelines leak. From February 2019 to March 2021, there were around 320 "oil- and gas-related incidents" ("oil spills, fires, blowouts and gas releases") just in the northwestern corner of New Mexico, mostly on Navajo Nation land. One such "incident" was the leakage of 1,400 barrels of fracking slurry into the local watershed.[63] Shall we now build more pipelines for the sake of the environment? This is yet another proposal sopping in climate reductionism, environmental racism, and coloniality. In the case of carbon pipelines, a leak would mean death by suffocation for all animal life forms nearby.

Beyond cane toads, we need to consider the very question of what we consider to be technology. After all, the Amazon rainforest constitutes a massive technology resulting from thousands of years of intelligent and reciprocal influencing by hundreds of native peoples. Next to them, carbon capture experts are amateurs. At this point it becomes necessary to distinguish between two very different paradigms of technology. What we can call interventionist or engineered technology is a Trojan horse for beliefs of human supremacy and practices of extractivism, exploitation, and depersonification of the Other against which the technology is deployed. Today, it is fully wrapped up in the legal regime of intellectual property and the institutional framework of the profit-oriented academy, laboratory, and research institute. It is a technology of enclosure, of the alienated worldview that looks down at our lives from above in order to impose a plan that necessarily divides governors and governed, designers and inert material. On the other hand, we have popular or traditional technology, which belongs to, is developed and passed down by a community; which is territorially specific, nuanced, and reciprocal; which constitutes a form of dialogue and commoning and therefore tends to be ecocentric; and which is necessarily anticolonial.

Half of any technology is deployment, how a specific artifact is integrated and mobilized throughout society in conjunction with other technologies and organizational techniques. Currently, interventionist technologies are deployed by a massive, ecocidal infrastructure and their use is controlled by corporations governed by the need to profit. The institutional paradigm reinforces technocratic, quantitative thinking that is reductionist and blind to complex realities on the ground, putting well-meaning scientists in bed with entrepreneurs and economists, giving rise to figures like the "carbon budget" that normalize the extreme suffering and damage that are already widespread.

To discuss BECCS or "rocks-for-crops" in the current context, as though once we invent these marvelous things we will be able to use them in a sensible manner, is to be completely oblivious of our surroundings. The extreme naïveté of brilliant scientists should never be underestimated. It was, after all, pacifists who gave us nuclear weapons (particularly Einstein, Meitner, and Szilard). The infrastructural and cultural context—who holds power and how power is shaped and conceived—matters. The intentions of inventors and researchers do not.

An acceleration of interventionist technologies will only accelerate the tendencies of power as it already exists. Already, technologies for controlling the weather are exacerbating pre-existing conflicts. Seeding clouds to capture rain or prevent it has pitted the owners of BMW factory lots in Mexico against local farmers, and contributed to the militarized conflict between China and India, with the latter unable to compete with the scale of the Chinese program to harness ever diminishing rainfall.[64]

As we will see in later interviews, the questions of what we mean by technology, who controls it, and within what power relation, on what side of the social war it is deployed, are all more important than the technical specifics of one innovation or another. In the meantime, I would suggest that those who cannot even go on record opposing intellectual property cannot be trusted to understand the dimensions of this problem. Every technology that is not fully under our control, fully expropriated from the capitalist market and colonial institutions, is another weapon pointed at us and at the planet.

RELIGIONS OF CONSUMPTION: ECOCIDE AND ENTITLEMENT

How mainstream media, politicians, and scientists speak about the ecological crisis often reveals a backpack laden with Eurocentric, colonialist myths. "In the beginning, human beings tended to view nature as a mortal enemy," according to an article in *The New York Times*.[65] Their examples come from the Old Testament and European colonizers of the Americas. In other words, while arguing for a scientific view of climate change, they embrace a Western mythological view of human beings, one that even seems to give a nod to the biblical notion that we've only been around a few thousand years. Discourses around human nature are almost always attempts to erase the *wrong* kind of human and to justify the status quo.

Western mythologies, for all their diversity and the breadth of their explanations, reach consensus on viewing humans as the superior life form, thus opening the door to hierarchies among humans between the more human and rational and the more chthonic or animalistic. In this consensus, nature is everything else, including and especially those humans who are not human enough. They all become resources to be exploited. In statist chronicles going back millennia, "wild" was generally a synonym for "stateless" and thus also for communities that practiced some form of commoning in a more reciprocal relationship with other living beings. (In early east Asian states, the metaphor frequently used was "raw," again connoting a more natural condition, not having passed through technological improvements.)[66]

In Genesis, God exhorts His followers to "fill the earth and subdue it, and have dominion over the fish of the sea and over the birds of the heavens and over every living thing that moves on the earth." Aristotle ranked all living beings on a ladder, with humans at the top, of course.

His ladder construct has remained popular among Western philosophers and scientists from the right, left, and center for the past two millennia.[67] Aristotle also believed the souls of women, children, and slaves were incomplete compared with the souls of Greek men. Cicero said that "all things in this world which men employ have been created and provided for the sake of men."[68]

Descartes argued that vivisection gave one a mechanical understanding of (other) animals, and that once one understood them mechanically, there was no reason not to kill them.[69] Mechanists influenced by Descartes' insistence that other animals had neither reason nor a soul promoted the idea that they did not experience emotion or suffering—a convenient idea for capitalism at that time, which was in full expansion and relying heavily on the labor power of beasts of burden and the first stirrings of industrialized meat production. This belief tended not to be shared by Descartes' lower-class contemporaries,[70] yet it remained prominent among scientists through the twentieth century. Inevitably, it was a belief and a practice that the foremost scientists among the slave-trading nations of Europe and in Nazi Germany applied to humans they considered to be less human.

People brought up in Western society—and to a differing extent and in more complicated ways, people brought up colonized by Western societies or in societies that follow Western economic models—are taught that the world is our gold mine and our outhouse. Even attempts to respect nonhuman life that base themselves in Western philosophy tend to be anthropocentric. The result is often a piecemeal approach that extends protection to the beings most similar to humans and the most aesthetically pleasing.

Plants and fungi are also capable of perception, learning, and making choices.[71] They have their own stories, their own value, but why stop there? It has become clear that cultures that have traditionally viewed the soil or water as living things are effective at preserving them, while the culture that treats soil and water as inert resources has created a planetary catastrophe in which most fertile soil has been destroyed and huge swathes of the globe are facing water shortages.

No healthy response to the catastrophe is possible without acknowledging that Western civilization is a failed civilization, and that its notions should be interrogated in light of its consequences.

Ideas of human superiority are not universal. Colonialism, again, is the culprit for silencing other views of our relation with the world, and it is largely from colonized peoples and struggles against colonialism that these other views are expressed.

Vietnamese Buddhist monk and teacher Thích Nhất Hạnh explains the connection between relations of empathy and knowledge that starkly contrasts with, for example, Descartes' practice of mechanism, apathy, exploitation, and objectivity. "When we want to understand something, we cannot just stand outside and observe it. We have to enter deeply into it and be one with it in order to really understand it."[72]

Sulak Sivaraksa, a Thai activist and teacher of Engaged Buddhism, writes:

> Even before the great traditions of Asia, animism contributed towards the cultural concepts of peace and social justice, encouraging respect for natural phenomena. We were taught to revere the spirits that look after the forests and oceans. At each meal, we expressed gratitude to the Rice Goddess to remind us not to eat wastefully and to be aware of all the human labor and natural resources that went into each plate of food. Traditional rites of the field also contributed to an awareness of and gratitude towards nature. Local festivals promoted communal spirit, reminding us that rice is for collective consumption rather than individual wealth.[73]

Niillas Somby, "a prominent figure in the Sámi's struggle for justice since the 1970s"[74] against the occupation and exploitive, assimilationist policies of the Norwegian, Swedish, Finnish, and Russian states explains how:

> You can see the same pattern everywhere colonialism took hold: traditional spirituality was the first element of the culture to be attacked … The land doesn't belong to us. We belong to the land, and it is our responsibility to take care of it.
>
> … Practicing traditional spirituality is not only about healing and medicine, it is about understanding how everything is connected, about being able to relate to nature and be its guardian.[75]

Sámi historian Aage Solbakk wrote, "Here in Sápmi, all of nature is alive. Everything has a soul, whether it is a birch or a rock, a lake or a creek."[76]

Mushkegowuk scholar Jacqueline Hookimaw-Witt defines "the difference between the Cree concept of land ownership and the colonial concept of private property" as follows:

> You cannot buy land and own it like you own a car. Yet, we do own the land in a different sense ... as much as we "own" the land, the other beings on the land (animals, plants, rocks) own the land as well, meaning that we were put on the land by the Creator, and everything on the land belongs there and can use the land.[77]

Métis and Cree writer Mike Gouldhawke goes on to explain:

> Land is the terrain upon which all our relations play out, and it can even be seen as a living thing itself, constantly shaping and being shaped by other life forms. Land isn't just a place, it's also a territory, which implies political, legal, and cultural relationships of jurisdiction and care.
>
> ... Traditional Cree laws like sihtoskâtowin (coming together in mutual support) and miyo-wîcêhtowin (the intentional cultivation of good relations) stand in stark contrast to this settler system, which is based on private and individualized rights to property and political representation.[78]

Kauy Bahe and Brandon Benallie explain the Diné concept of *k'é*, which anthropologists catalogue as a kinship system. It is much more than that, says Bahe, a "huge overlapping philosophy that the whole universe is interconnected. But it's also these relationships that we have with one another and with the elements that exist in the world, whether that be the weather or the water or the animals." Benallie calls *k'é* "our theory of everything ... It's how we're connected to everything—but specifically how that kinship is reciprocated and maintained."[79]

Dr. Ku'ulei Rodgers, a Hawai'ian coral reef ecologist, explains how the Kumulipo, the Hawai'ian creation story "talks about how the first two beings bring forth organisms in the order of importance. The first is the coral polyp, followed by the coral colony, then the marine creatures, then the terrestrials. The Native Hawaiians understood that their connection to the sea was absolutely vital to their lives."

Christian missionaries like Daniel Dole, whose family would go on to create the world's largest fruit canning company, came to Hawai'i in the mid-nineteenth century to destroy the Indigenous belief in a spiritual connection with the sea and the earth, while also snatching up huge tracts of land for business ventures. Restoring the traditional relationship and the communal fishing practices that go along with it is vital to saving Hawai'i's endangered marine ecosystem.[80]

Standing Rock Sioux scholar Vine Deloria Jr. found that "Land acquisition and missionary work always went hand in hand in American history" and that "Exploitation by one's fellows became a religious exercise. Law became a trap for the unwary and a dangerous weapon in the hands of those who understood how to use it."[81]

By rejecting Western civilization, I am most specifically saying that we need to eradicate colonialism, which is really the inaugural event that created the pernicious concept of the West. We need to eradicate all the political and economic institutions that carried it out, and the philosophies that gave it an alibi, such as the nested ideas of human and European supremacy. I recognize that hardly any social structure or belief form on the planet has been untouched by colonialism, and that many of us, myself included, have been brought up completely within the matrix of structures, practices, and beliefs that make up the nebulous, fictive, and always poorly defined West. We can't just choose a new culture and a new history as though such things were items to be found in a supermarket; at least, we cannot do so while fighting colonialism. I recognize that we are faced with a world of ruins, and that every community will have to sort through these ruins and decide which ones they treasure, which ones to throw on the bonfire, and which ones can be repurposed for a healthy world based on solidarity and reciprocity.

At a minimum, this critique means recognizing and refuting the ideas that are hegemonic in present society, the ideologies we often do not recognize as such and that all the dominant institutions reproduce. This includes capitalism, which has wormed its way into environmental movements with the concept of "sustainable development." Despite numerous studies that continued economic growth and ecological sustainability are not compatible, that green growth is simply not possible,[82] generations of pragmatic environmentalists and ecologically minded scientists have been duped into dedicating their efforts to the chimera of green growth. As argued by Demaria, Kallis, and Bakker, "growth is not only a material and economic process with social and ecological costs, but also a hegemonic idea that obscures more ecologically friendly and egalitarian alternatives."[83]

A critique of this failed civilization also extends to the dominant practices and institutions of science. And it is, to put it lightly, impolitic to criticize science at the current juncture, because of the prominence of climate denialists willing to dismiss an overwhelming consensus regarding the ongoing ecocide that threatens us all. However, this is not a clash between irrational, superstitious people and people of science, as progressive media would have us believe. In fact, the very trope of superstitious hordes is both racist and inaccurate, as the so-called tribal peoples evoked by the cliché, insofar as it ever referred to real people, were people who spoke about the dangers of destroying the environment long before scientists did, and their spiritual practices—the much-maligned superstitions—were a key element of their intelligent, ecocentric forms of life.

The sad fact is that science denialism is a fully scientific phenomenon, fed by the advanced algorithms of Facebook, the many scientists in the employ of the fossil fuel industry, and the richly financed studies on what kind of populist lies politicians can get away with.

The problem with the "believe science" slogan is that it presents scientific research as a neutral, harmonious process. Critiquing science as it exists today—the web of institutions, funding and employment opportunities, and research agendas inseparable from the cultural priorities and prejudices of those who design them—is not a rejection of empirical knowledge production or any conceivable iteration of scientific institutions that might exist in alternate universes. It is a way to unmask the mythology that presents science as a pure synonym for knowledge and to talk about what is actually happening right now, and what needs to happen.

Ecocide is a highly technical process. Not a single one of the industries responsible for destroying our home has developed without the integral participation of academically trained experts. Not a single oil well, not a single gold mine, not a single fracking site. Scientists are shaped by the same mercenary pressures as all higher value laborers in a capitalist economy.

The scientific academies and institutions that exist today generally arose or matured as part of the global process of colonialism. The Royal Society in the UK, the world's oldest scientific institution, was an active accomplice in the military and economic expansion of the British Empire.[84] The most prestigious US universities exist in large part to support—and derive much of their funding from—US militarism. Disciplines from geology to anthropology arose as part of the colonial project, and within extractivist mining industries or counterinsurgency operations in Iraq and Afghanistan they continue to play that role.[85] Botany and Chemistry, together with the pharmaceutical industry, simultaneously belittle and exploit Indigenous knowledge of local plants and ecosystems.[86] Kevin Chang, a Hawai'ian activist with Kua'āina Ulu 'Auamo, an organization that pressures the government to allow community stewardship of the land and waters, relates how: "conservation work was tied solely to a western way of looking at biology and ecosystems. In the past outside 'experts' weren't connected to the communities they studied. It's forgotten that expertise exists in the community itself."[87]

Among the people I interviewed for this book, participants in ecological and anticolonial struggles around the world, a distrust and hostility towards academics came up again and again, unprompted. They were frequently identified, alongside NGO workers, celebrity activists, and politicians, as those who exploit and even help destroy social movements in line with the interests of those who fund their institutions.

Communities in resistance could make great use of the resources academics have to offer, and in some examples we'll look at in Chapter 3, they already are. For this to be possible, academics need to learn how to listen: to the land, to communities, to people in struggle.

Whether this changes comes down to an important question of positionality: where will they locate themselves in relation to the powerful institutions of knowledge production that molded them and the communities those institutions seek to extract knowledge from? Will they come to communities in struggle as saboteurs and thieves, ready to dynamite the enclosures their bosses have built around common knowledge, stealing from their institutions and giving back to the community, helping knowledge and learning become collective properties again? Or will they come as they were designed to, like factories or transatlantic ships, bearing false promises of progress, plundering and disciplining, hoarding the benefits and externalizing the costs?

Simply put: it's up to them.

However, we cannot wait for them any longer. The experts who caused the problem are not going to solve it, certainly not without acknowledging their extensive institutional power and clearly positioning themselves on the sides of communities in struggle and against their funders, against their own careers.

In the meantime, people are doing all they can to collectivize knowledge. This also means recognizing and harmonizing two different kinds: knowledge that is quantitative, falsifiable, and universal; and knowledge that is situated, corporeal, historical, and self-aware.

To place this critique within a living history of resistance, I spoke with Adrianna Quena. Adrianna has participated in the food and agricultural resistance movements in Venezuela of the past two decades, "which in turn are heirs, of course, of much longer historical processes of struggle for the land and sovereignty in Venezuela." She has been a part of collectives and platforms like la Campaña Venezuela Libre de Transgénicos (Campaign for a Venezuela Free of Transgenics) and el Movimiento Popular Semillas del Pueblo (Popular Movement Seeds of the People). More recently, she has migrated to Barcelona, where she takes part in the urban garden Quirhort and La Cuina, "an artistic laboratory working on food and politics."

> I believe history has shown us enough times that "formal" or academic science is not neutral. It is configured in such a way that the knowledge produced in turn reproduces colonial and patriarchal logics that expand the domination of the scientific institution it represents. It is a Eurocentric institution that dances to the tune of capital. There are exceptions, but I believe technocratic propaganda already gives them plenty of coverage and we try to speak from a different place.
>
> Does that mean that knowledge and consciousness are affairs of the powerful? No and No. It means that in the notion of the "expert" there is always a latent racism, a latent authoritarianism that is quite dangerous, that legitimates itself as something superior, that underestimates and/or commodifies popular scientific knowledge as well as the people who produced it and those who have tended to it, as a network and across many generations.
>
> If what we need to survive on this planet is to repair our ecosystemic relations, and we understand that the academy, the universities, the research centers, are colonial institutions, and we also understand that colonization is at its base an ecological problem, with its criminal logic of accumulation by dispossession, and that the problem is differentiated between the North and the South, between the city and the countryside, I think it is worth stopping to think: what kinds of questions can be posed from within that academic, scientific institution? And we should also ask, or at least doubt: how is it that the knowledge that arises exclusively from these institutions is more useful or more legitimate that the knowledge produced by people in their own territories, that have been wrapped up in absolute devastation but also intense climate resistance for years now?

I think that the scientistic approximation to knowledge damages territories as well as the comprehension of the world that is possible, because it is too fragmented and because it homogenizes, tries to standardize solutions and in so doing it loses—and sometimes annihilates—the adaptive and sustainable knowledge that has made human life possible on this planet for thousands of years.

As a semi-urban, white-passing granddaughter of displaced peasants, I could feel how my path through the academy offered me, basically, the construction of a supremacist subjectivity, with important degrees of self-hatred, that did not allow me to even formulate coherent questions regarding the history of despoliation of our territories in Venezuela, nor to understand the needs and desires that beat in our hearts as a result of this history, of these resistances. The mere attempt to formulate those questions was considered subversion and, as a result, persecuted.

Once I had already learned this, I decided to study for real, and I shifted my gaze towards peasant and Indigenous knowledge and I realized that (1) I am completely ignorant and (2) the true science, which is to say observation, experimentation, innovation, and the transmission of popular knowledge and technology in Latin America and, particularly, in Venezuela, has resisted 100 years of modern petroleum-linked destruction and 500 years of colonial devastation.

As I learned—listening, caring and letting myself be cared for, sowing, navigating, singing, and cooking—I witnessed the arrival of technicians and experts of every type: engineers, anthropologists, agronomists, political scientists, bureaucrats, pedagogues, philanthropists, and activists ... and I saw a thousand times how that superior gaze made collaboration and a complementarity between knowledges impossible. Other ways of relating and producing knowledge are totally possible, but they don't teach them in university or in the government ministry.

Global ecological problems like hunger, the loss of agro-biodiversity (from 10,000 edible species current diets only deal in about 150), soil degradation, the slow death of the seas, the displacement of millions of people, victims of resource wars and so on, requires more horizon, more commitment, and more risk than some academic degrees or the completion of a government audit obeying some arbitrary and decontextualized limits.

To sum up, we could say that the logic of the expert makes it impossible to imagine this other culture that we must urgently create, and it makes it impossible to weave an ecology of reparations that we need to survive.

On the other hand, also to sum up, we could say that the logic of knowledge as a network, adaptive and not commodified, is the most important beacon to orient ourselves and make sure the future exists. What can we learn from this knowledge? For me the greatest lesson is that quality is the most important and sustainable thing. A territory's criterion of quality gathers together the ethics of that territory's community, its notion of what is life, what is justice, what is abundance, and what is wellbeing.

PRISON CELLS AND DEATH SQUADS: THE REAL FIRST RESPONSE TO CLIMATE CHANGE

In April of 2009, representatives of the 28 member states of the North Atlantic Treaty Organization (NATO) met in Strasbourg, France, and just across the border in Kehl, Germany. The largest military alliance on the planet had climate change on its agenda at a time when politicians in most member states were at best inactive and at worst actively denying the reality of the situation. What solution did they propose to climate change? Increasing "border security" as well as biometric IDs and more surveillance for their domestic populations, and reemphasizing their goal of normalizing the domestic use of military forces in urban environments by 2020. A strategy paper released in advance of the summit said plainly that "Security challenges are predominantly socio-economic, not military-technical, in character."[88] The paper goes on to make clear that "socio-economic" threats are those posed by poor people, climate refugees, and others angered or harmed by the current state of affairs, and as such, member states should continue to integrate police and military action and be prepared to deploy the military in their home territories.[89]

Anyone who is shocked by this response, or sees it as some kind of *non sequitur*, has failed to appreciate the true nature or scope of the ecological crisis. World governments, particularly those in the Global North, understand the crisis as a security issue. They know the problem is real, and they know growing deserts and rising sea levels will force hundreds of millions of people from their homes in search of their very survival.

Their solution is to further militarize the borders—those borders of Fortress Europe and the American southwest that are most definitely "designed to kill"—so that people fleeing for their lives will be dissuaded by the very real possibility of dying in their journey.[90]

To put it simply, the major militaries of the world, already among the greatest producers of toxic waste and greenhouse gases, propose killing even more people to shield themselves from the consequences of the crisis they are in large part responsible for. And they also propose increasing the repression against their own citizens, fully aware that the lower classes everywhere will bear the brunt of the crisis, and rebellions are to be expected in the North as well as in the South.

While the rich and powerful rubbed shoulders at the summit, thousands of anarchists and other anticapitalists took to the streets, battling with police and burning down banks as they tried to disrupt the NATO summit. Though the media condescendingly portrayed them as a mindless horde bent on evil deeds, they were perfectly aware of what NATO was proposing and the stakes for all the rest of us. Their response might be characterized as the most reasonable and intelligent, at the very least if we compare it to that of the scientists and NGOs who continue to dialogue with the same governments sponsoring the summit.

Three kinds of people with the most experience and the most effectiveness at resisting the ecological crisis have been systematically excluded from the official conversation on solutions to climate change: Indigenous peoples trying to protect or recover their land and their traditional forms of life; people most affected by the ecological crisis, including climate refugees; and radical environmentalists, radical in the true sense of going to the root of the problem, and therefore, typically, participants in anticapitalist and anticolonial movements.

These should not be thought of as three separate groups. Many Indigenous people are anarchists and socialists; many migrants, or activists opposing incinerators or coal mines in their neighborhoods, are Indigenous, or anticapitalists, or both. Instead, we can think of these as three different frontlines that intersect on countless occasions.

It is significant that the state treats people on all three of these lines as social enemies to be surveilled, as threats to be eliminated. NATO is not exceptional. From Borneo to Brighton, when people start taking action to defend the Earth and fight for a dignified survival, the default response of governments and the economic interests they work with hand in hand is to resort to death squads or prison cells.

On Borneo, one of the main threats to the environment, to Indigenous communities, and to poor people stuck between small farming in an inhospitable economy and forced urbanization with unhealthy conditions, is deforestation carried out primarily in connection with the palm oil, logging, and paper pulp industries which constitute the main sources of revenue for the local oligarchy. Deforestation causes deadly air pollution, destroys soil, and spells an end to the small-scale farming many communities depend on. Those who resist face violence from police or paramilitary thugs working on behalf of the oligarchy.

One activist from Borneo who works on agrarian issues and Indigenous land struggles, particularly with several Dayak communities, told me how the state neutralizes resistance. "In the worst cases, the state murders its opponents." Though that happened more frequently during the New Order era, after a US-backed dictatorship murdered over 500,000 people, it still occurs today. The state also "criminalizes and arrests Dayak activists or cultivators ... to cover up the actions of the palm oil companies ... Sometimes, arrested activists die in police custody." The activist described one case of a dissident journalist who was tortured and mutilated after writing critically of the business oligarchy that controls politics on the island. The journalist subsequently died in questionable circumstances. "I could name other cases of repression, but it's a bit psychologically exhausting."

During the week that I was working on this section of the book, two Indigenous land defenders were killed in Honduras. José Adan Medina and Félix Vazquez, of the Tolupan and Lenca peoples, respectively, were assassinated for opposing hydroelectric dam construction and other land abuses.[91]

According to the NGO Global Witness, between 2002 and 2018, over 1,700 land defenders—people trying to stop the destruction of the environment or human rights abuses related to resource extraction—were murdered in 50 countries. Some 40 percent of the dead were Indigenous and the killers, the vast majority of whom committed their deeds with impunity, were police, military, and paramilitaries working at the direction of the state or of multinational companies.[92] A record number of land defenders were murdered in 2019, with 212 deaths recorded and many more unrecorded.[93] Colombia, the Philippines, and Brazil topped the list for murders (with the caveat that extractivist violence in many African countries is underreported, in part for reasons we will look at shortly).

In Brazil and especially in Colombia, paramilitary death squads created with the support of the United States attack communities that stand in the way of the large economic interests that destroy the rainforest to establish mines, logging operations, and plantations.

In Colombia, Dole Food Company, Chiquita Brands International, and other companies paid paramilitaries several million dollars to protect their interests, including killing union organizers among their workers. The companies would pass the names of labor organizers, community activists, and suspected leftists to the paramilitaries, who would then execute them.[94] Between 1997 and 2004, paramilitaries committed 4,335 homicides and 1,306 forced disappearances in communities near Chiquita plantations.[95] And while it was on the Chiquita payroll, the AUC paramilitary group forcibly evicted 60,000 people from their homes in the banana growing region of just one state. Even after the group was declared a terrorist organization by the US State Department (a move that came only after decades of public pressure, given US government complicity with the paramilitaries), Chiquita was sending them money and smuggling in thousands of assault rifles for the AUC using the company's international distribution facilities.[96] They were punished with a small fine after a 2007 legal action, in which they were protected from serious consequences by attorney Eric Holder, whom President Obama would soon appoint as Attorney General.

Throughout the Colombian civil war that began in 1964 in large part due to the United States' Cold War policies, 262,000 people have been killed, 82 percent of them civilians, and 120,000 are still missing. The lion's share of confirmed civilian killings were carried out by paramilitary death squads (100,000), even though mainstream media citing the same study I am presented the news in a way that suggests leftist guerrillas were responsible for more killings (35,000).[97] The Colombian military, which received billions in aid, training, and weapons from the US, also had plenty of blood on its hands. Killings by the military became so systematic, officers began establishing quotas and offering cash rewards for higher body counts, leading to soldiers executing more than 10,000 civilians just between 2002 and 2010—largely homeless people or people with mental disabilities they had kidnapped or lured away from the cities—they falsely identified as leftist guerrillas in order to increase their kill counts and claim the rewards.[98]

In 1995, Ogoni writer and activist Ken Saro-Wiwa was executed by the Nigerian state, together with eight other activists who had been nonviolently protesting the destruction of Ogoni lands in the Niger Delta by the oil industry, particularly Royal Dutch Shell. His last words were, "Lord take my soul, but the struggle continues." He was right. In December 1998, a youth conference of the neighboring Ijaw people committed to a campaign of direct action against the oil industry, albeit nonviolently. The Nigerian military invaded the region with over ten thousand troops and opened fire on a protest march with machine guns, killing at least three. They again opened fire on a march demanding the release of those arrested, killing at least three more, and then invaded communities, terrorizing the people and raping women and girls. In January 1999, a hundred soldiers directly attached to a Chevron facility attacked two Ijaw villages, killing and disappearing dozens.[99] Despite this extreme level of violence, the Ijaw did not give up, and one might say they could not give up.

The oil industry makes their lives impossible. Frequent leaks and explosions poison fields, destroy forests, rivers, and coastal waters, and kill workers and neighboring villagers. Since Dutch, British, and US companies began oil extraction in 1956, they have leaked one and a half million tons of oil into the land, or a current rate of a quarter of a million barrels every year.[100] Increasingly, local Ijaw and Ogoni people cannot feed themselves from the fields, forests, and fisheries on which they had traditionally depended. This is a perfectly acceptable state of affairs within the sort of development framework favored by the World Bank and similar institutions: people *should not* feed themselves from the land. Subsistence agriculture is dismissed as a primitive activity. Rather, they should get jobs and then pay for food from industrialized and globalized producers (relying, of course, on oil to power their machinery, manufacture their chemicals, and transport their product). However, this is also not an option for the peoples of the Niger Delta, as all the profits of oil extraction go to foreign companies and the central government.

The Ijaw took up arms to defend their lives and formed groups like the Movement for the Emancipation of the Niger Delta. MEND sabotaged oil facilities and kidnapped highly paid North American and European oil workers. Their actions and those of similar militant groups led to major reductions in Nigerian oil production, for example shutting down 10 percent of production with a single attack in 2008[101] and leading to widespread reductions over the better part of a decade. The Nigerian military responded with bloody repression, including aerial bombardments of Ijaw communities. The US military also stepped in to help the government better defend its oil platforms in the name of fighting piracy,[102] helping oil production in the Niger Delta to grow again.

The state-sanctioned murder of activists using what the West deems to be legitimate (nonviolent) tactics to defend their lands and their livelihoods frequently drives their communities to take up arms just to survive. Subsequently, they often become more able to attack and stop the companies responsible for poisoning them. However, when they are then massacred by their governments with weapons from Europe and North America, human rights NGOs no longer count their deaths—or those of the community members supporting them and giving them shelter—in their lists of land defenders and environmental activists killed each year. So, in fact the above-cited figure of 1,700 murdered land defenders is woefully inadequate.

The violence is by no means a phenomenon of the Global South. It is the result of the colonial, capitalist relations that transect the entire world. The profiteers and the shot-callers can be found wherever there are major concentrations of political and economic power, from Brasilia to London. Contrary to the illusions of civility that are central to every democracy, the violence, while unequal, is also global. As the European Union promotes itself as a model for the green capitalist future, it also profits off the same, gory repression. In the last few years, seven Romanian park rangers have been murdered and over 650 have been assaulted or threatened for going up against logging interests connected to an Austrian company, with the line between legal and illegal logging "increasingly blurred."[103] And environmental activists have been threatened or have had to go into hiding due to attempts on their lives after they investigated logging, hydroelectric dams, and other industries. Incidentally, these hydroelectric dams and these forests-turned-logging-plantations are the ostensible carbon sinks and nature preserves that lie at the heart of the European Union's model for profitable environmental action.

In North America and Europe, the government classifies people opposing pollution, deforestation, and global warming as "extremists" and treats them accordingly. In the US, the FBI and Department of Homeland Security listed "racial and environmentally themed ideologies" as primary motivators for domestic terrorist attacks. As for those "racial" ideologies, the FBI focused a great deal of attention on so-called Black Identity Extremists, its way of discrediting those who are outraged by racist police murders. The property damage of antiracists and environmentalists merited such attention while white supremacists were carrying out hundreds of attacks, many of them lethal, including full scale massacres. Activists who carried out civil disobedience against oil pipelines—simply turning off valves—were placed on the same "extremism" lists as those responsible for mass shootings in Black churches.[104]

As Andrea Brock has documented, Europe is no stranger to the same dynamics. Those who set up encampments to defend forests are treated as enemies of society and met with militarized repression and marginalization by interlocking consortia of public and private security, elected officials and corporate board members.[105] As we shall see in the next chapter, farmers and anticapitalists occupying the land and opposing the construction of an airport at Notre-Dame-des-Landes, France, were repeatedly attacked by a militarized police force, with one young person having their hand blown off by a flash bang grenade. German police defending the expansion of an expanding coal mine that is destroying the Hambach Forest regularly brutalize protesters. Numbering in the thousands and backed by military grade weaponry, they have cut protesters down from six-meter-high tree sits, shot them with high pressure water cannons in below freezing temperatures, struck people with vehicles, attacked them with dogs, and kicked their teeth out.[106]

Italian police shot anarchist Carlo Giuliani in the head, killing him, during heavy protests against the G8 in Genoa in 2001, where they also raided a protest convergence point, brutally assaulting and torturing dozens of people trapped there. At the G8 protest in Evian, France, in 2003, police cut down an activist suspended from a bridge, causing him to fall twenty meters and almost killing him. In 2014, French police killed an environmental activist during heavy repression against a campaign to stop deforestation associated with a new dam. The campaign was eventually successful, and the dam was canceled.

At anticapitalist and ecological protests across North America and Europe, police routinely attack protesters with less lethal weaponry, regardless of whether they are being peaceful or combative, occasionally causing deaths and more frequently leaving people permanently injured. Hundreds of people have had eyeballs shot out or suffered brain damage from high impact munitions. At the protests against the Free Trade Area of the Americas in Miami in 2003—a major neoliberal initiative that would have accelerated ecocidal practices across both continents and which was defeated thanks in part to fierce opposition from the streets—

police raped detainees. Sexual assault by police during arrest, especially against women, nonbinary, and trans people, is common.

To preserve the illusion of democracy and the beliefs of privileged citizenry that the institutions of power are on their side, police forces in the Global North use a higher proportion of "soft" counterinsurgency techniques, but these come with their own forms of violence. Many participants in the George Floyd Rebellion in 2020 associated counterinsurgency with discourses of reformism and nonviolence that facilitated the isolation of participants in the uprising, exposing them to the violence of the police, National Guard, and prison system, all of which could be naturalized if they were seen as being used against a criminal minority and not against the movement itself.[107] This interplay illustrates how soft and hard counterinsurgency methods go hand in glove.

Effective soft counterinsurgency requires massive intelligence gathering. Spying on politically active people in the northern democracies, even to an extent that violates the applicable laws, is systematic. This includes the long-term infiltration of anticapitalist movements. In the UK, around 150 police agents went on deep infiltration assignments against anticapitalist, environmental, and other leftwing movements since 1968, living under assumed identities and participating in those movements for several or even a dozen years. Agents routinely began romantic and sexual relationships with activists they were informing on, using the forms of manipulation they had been trained in. Several of these undercover cops impregnated activists, committed to forming a family with them, and when their stint was up, disappeared.[108]

Though the details that have become public are horrible, we should assume the truth is far worse: police were caught destroying a large quantity of documents they had been ordered to preserve concerning the infiltration program since the 90s, when it expanded and shifted largely to environmental and anarchist groups.[109] Since the scandal broke in the news, the UK has introduced a new law to officially allow domestic police to carry out illegal acts while undercover, with support from both major parties.

Between 2010 and 2014, the Spanish state directed its substantial antiterrorism apparatus against a new public enemy, anarchists, carrying out raids in a half dozen cities, smashing down doors, waking people up with guns in the face, stealing all electronics, disappearing all cash, and hauling more than fifty people off to jail.[110] In those years, anarchists had increasingly and effectively taken part in popular struggles against austerity measures, evictions, and deportations. They had also gotten the government's attention with animal liberation actions and campaigns against the MAT, a high-tension power line from France to Morocco connecting energy markets and extending the distribution of nuclear, coal, and green energy.

In the US, the use of antiterrorism against ecological struggles can be traced back to the '90s, when the logging industry popularized the concept of ecoterrorism amidst the fight to save the redwoods. With massive resources, PR firms, and connections to the media, the logging industry tried to convince the public that those who broke laws and damaged inanimate property in defense of the Earth were terrorists and a threat to society.

Hollywood responded to the needs of industry with an enthusiastic entertainment/propaganda drive that undermines US society's self-image of having a free media. Since the '90s, over a hundred movies and TV shows have been cranked out to summon the specter of ecoterrorism, featuring bad guys who wanted to kill, cause mayhem, or destroy the entire human species for vague environmentalist motives. These works of industry propaganda span the spectrum from comedies to Michael Crichton thrillers, James Bond adventures to crime shows. On the contrary, very few movies and shows feature people suffering from environmental racism, and even fewer show people breaking the law as an act of conscience or necessary resistance. Protagonists in environmentally themed movies are nearly always white lawyers, journalists, or citizen activists like the protagonists of *Dark Waters* and *Erin Brockovich*.

Industry lobbying paid off. In 2002, shortly after the attacks of September 11, 2001, and in the midst of the new War on Terror, the US government declared radical environmentalists to be the number one domestic terrorism priority, at a time when lethal white supremacist attacks were already on the rise and groups like Earth First!, the Earth Liberation Front (ELF) and the Animal Liberation Front (ALF) were not even accused of having committed bodily harm.[111]

In the Green Scare, the US government brought its full repressive weight against these movements. Jeff "Free" Luers and Craig Marshall were arrested in June 2000 for setting fire to three SUVs. Luers was given a 22-year prison sentence, later reduced to 10 years only thanks to a well-organized support campaign. As part of Operation Backfire, in December 2005, the FBI indicted 13 alleged members of the ELF for dozens of arson and sabotage attacks against the Vail Ski Resort, power lines, genetic engineering research facilities, luxury housing developments, and other targets. They were threatened with hundreds of years in prison. One detainee, William "Avalon" Rogers, committed suicide shortly after arrest. Most non-cooperating defendants ended up with prison sentences of six to seven years only because the FBI had to offer plea deals to avoid revealing what were probably illegal surveillance tactics should the cases go to trial.

In 2006, Eric McDavid and two others were arrested on conspiracy to bomb a dam as part of a plot that was funded, organized, and pushed ahead by a paid FBI informant who had become romantically involved with McDavid as part of her infiltration activities. Initially, McDavid was sentenced to 20 years in prison, but had his sentence reduced when the defense discovered that the FBI had withheld thousands of pages of evidence regarding the role of their informant. Also in 2006, six members of Stop Huntingdon Animal Cruelty, a group that coordinates aggressive boycott actions against Huntingdon Life Sciences, one of the largest animal testing companies in the world, were sentenced to three to six years in prison for sending large numbers of faxes to HLS or "internet stalking." They were convicted under the Animal Enterprise Terrorism Act, a law that allows the government to grant special punitive protections to companies that abuse and exploit animals.[112] A year later police in the UK, Netherlands, and Belgium coordinated a series of raids and arrested 32 people connected to SHAC. When HLS was on the verge of bankruptcy because the pressure campaign had effectively driven off other companies from doing business with them, the UK government stepped in to help the company secure financing.[113]

In 2009, Marius Mason was sentenced to 20 years in prison for an arson attack against a university lab involved in genetic engineering research funded by Monsanto, arson of logging trucks, and attacks against the construction of luxury homes. Like other Green Scare prisoners before him, Mason is being held in a Communications Management Unit, an opaque corner of the federal prison system that uses extreme isolation and a prohibition on communication with the outside world as a form of psychological torture. CMUs are used almost exclusively against prisoners of the War on Terror, anarchists, and radical environmentalists.

This policy of treating land defenders as terrorists has not ended. At the end of 2020, two people in Washington state were given federal terrorism charges for blocking a rail line as part of a campaign against the construction of new gas pipelines.[114]

Corporate involvement in such repression also continues. It was recently revealed that agrochemical giant Monsanto runs an "intelligence fusion center" to compile information on and conduct disinformation and harassment campaigns against journalists and activists who threaten the company's financial interests through their research or organizing. "Fusion center" is the same term the FBI uses for its counterterrorism centers. In just one example, Monsanto targeted a Reuters journalist investigating the carcinogenic effects of the company's star product, glyphosate, or Roundup. Their campaign included coordinating "third parties" to post negative reviews of the book, hiring scientists to cast doubt on the book's conclusions, pressuring the journalist's editors at Reuters "very strongly every chance we get" in the hope "she gets reassigned," covering up their financial relationship with scientists claiming their product was safe, accusing the journalist of being a "pro-organic capitalist" activist, as though there were big bucks to be made in opposing some of the world's largest chemical companies, and contracting search engine optimization (SEO) experts to make sure that their alternative facts, their negative reviews, and their various slanders of said journalist would appear in search engines above results showing how Roundup causes cancer.[115]

The above case illustrates how corporations can orchestrate subtle campaigns of censorship, often without revealing their hand. In 2020, an academic publisher abruptly canceled the publication of a book that showed how Canadian mining companies benefited from the genocide in Guatemala, moving in to stake their claims sometimes even before the death squads had left. The publishers expressed fears of lawsuits for defamation, though they refused to point out what part of the book, which received favorable peer reviews, might be considered defamation.[116] And in Canada, the RCMP spied on the release event of a book against mining.[117]

The new media environment is also particularly hostile for people trying to defend the land. Facebook is largely responsible for the rapid growth in the extreme Right around the world, encouraging racist conspiracy theories and climate denialism. Beyond the algorithms, when it comes time for administrators to block content, they rarely ban far-right groups, even after such groups carry out shootings and stabbings, but are very quick to block the anarchists and other antifascists who carry out the research and reporting necessary to protect our communities from neo-Nazis. When Facebook finally agreed to tamp down on misinformation related to climate change denial, they actually gave the boot to Indigenous and environmentalist pages that were fighting climate change.[118]

Direct police interventions also make the internet an unequal terrain for movements fighting for liberation. Articles I used in my research, particularly two texts by Indigenous anarchist Tawinikay that I rely on extensively in Chapter 5, became unavailable after Dutch police raided and seized computer servers for several anticapitalist news sites.[119] For us, this is a common occurrence, almost completely erased from the mainstream, where right-wing pundits and transphobic authors, all of them millionaires with enormous platforms, control the debate around free speech and complain of being "canceled."

Even progressive media lend their support to policing operations against environmental movements. In a typical article, the *Guardian* accuses anarchists of trying to "hijack" a climate march, the classical maneuver of dividing the resistance into "good protesters" and "bad protesters."[120] What their *outside agitator* caricature leaves out is that anarchists have been a part of the struggle to save the planet since way before the *Guardian* began dedicating ink to hijacking the movement to favor institutional perspectives.

The state deals with migrants with the same apparatus it brings to bear against social movements. Intelligence gathering, surveillance, demonization and scapegoating in the media, imprisonment, torture, targeted killings, and not least of which, depoliticization. Just as those fighting for clean water, a healthy relationship with the land, or a livable planet for future generations are deemed extremists, people who go through the trauma of leaving their homes behind are denied all reason and presented as criminals.

The economies of the US, Canada, and the European Union depend on immigrant labor. As Harsha Walia documents, the purpose of the border regime is not to stop immigration but to control and terrorize it.[121] When Germany decided in 2015 to take in over a million refugees from the Syrian civil war, it was only because the largest association of German employers had just declared that the national economy faced a shortfall of millions of skilled laborers.[122] But at no point did the German government allow direct flights from Turkey or Lebanon, where so many of the refugees were warehoused. Instead, they obligated refugees to make the expensive and perilous journey over the Aegean Sea, through the Balkans, under and over razor wire fences, through police truncheons and tear gas, past violent, xenophobic crowds, so that finally they would arrive with almost nothing, willing to accept any labor conditions and bureaucratic controls. It was a journey that cost on average several thousand euros, on top of the steep psychological price, effectively ensuring that primarily only members of the university-educated middle class would be able to make it.

Migrants from Africa are made to cross the open sea on flimsy rafts, left in the water or even deliberately rammed by the coast guards of the various Mediterranean states: between 2014 and early 2020, over 20,000 have died in the crossing.[123] Once across, hundreds of thousands of migrants are either held in detention centers—sometimes for years—or left to live in crowded shanty towns, often without access to running water, heating, toilets, or medical facilities and at the mercy of arson attacks by fascists often connected to the police and operating with impunity, from Spain to Greece.[124]

In the US, the border regime has been designed to force migrants to cross in the most dangerous regions of the Sonoran Desert. Between 1998 and 2019, Border Patrol recovered the bodies of 7,800 migrants who died while crossing.[125] Many more have died on the Mexican side of the border, or crossing through Guatemala and southern Mexico, where the Mexican government has increased enforcement in line with US demands. Border control, one of the branches of police with the strongest far Right sympathies, often brutalize and occasionally kill those who are crossing, as do rightwing militias. They also destroy caches of water left in the desert by volunteers trying to decrease the number of deaths. In detention, people are held in cells or cages with temperatures near freezing. People are left to die of medical neglect, or denied attention in the case of pregnancies, leading to frequent miscarriages; hundreds and possibly thousands have been involuntarily sterilized, leading some prisoners to compare the centers to "an experimental concentration camp."[126] Torture is systematic and "inherent" in the US immigrant detention system, according to a new report, and many are brutalized and forced to sign their own deportation papers, even though it means going back to a country where their lives are at risk.[127]

Australia is another terrible case of institutionalized racism that reveal borders to be an active measure of ongoing colonialism. The country legalized the buying and selling of enslaved Aboriginal laborers into the 1970s, and had an immigration policy explicitly designed to only allow white immigrants, also lasting until the 1970s. In 2018, the government was forced to pass a new law outlawing slavery again after finally acknowledging the tens of thousands of Pacific Islanders working in the agricultural sector, getting paid only $10 a week after deductions by their employers.[128] Determined to keep out "unskilled" immigrants from non-white countries, the Australian government obliges its poorer neighbors like Papua New Guinea to imprison asylum seekers indefinitely in facilities that have been compared to concentration camps.[129]

Meanwhile, solidarity with migrants is being increasingly criminalized. While humanitarian aid constitutes one of the mechanisms by which migrants are surveilled, controlled, and maintained in abject precarity and poverty, direct action and solidarity has helped migrants win housing and gain resources on their own terms, or pass clandestinely through strict border regimes. In Europe, people are increasingly being legally required to report on undocumented people and prohibited from giving them food or water; squats and autonomous spaces created for and by migrants are targeted for eviction; autonomous transportation infrastructure that provides mobility to undocumented people is criminalized and shut down; and those aiding migrants have been charged with trafficking, which carries a sentence of up to ten years imprisonment. Ironically, European governments and their private partners, who cynically paint migrants and border activists as traffickers, have funneled millions of euros to the Libyan militias that actually control both the imprisonment and the trafficking of refugees at a key point on the southern Mediterranean coast.[130] In the US, people can be imprisoned for years for giving migrants a map or giving them a ride to the hospital, and Border Patrol has increasingly been raiding a first aid camp set up by volunteer group No More Deaths. Migrants who speak out about abuses frequently get targeted with speedy deportations.

In July 2019, as the Trump administration was gearing up to carry out major raids targeting undocumented people across the country, anarchist Willem Van Spronsen set fire to an ICE vehicle at the Northwest Detention Center in Tacoma, Washington. His action and other widespread acts of resistance succeeded in getting ICE to dramatically scale back the raids, but it came with a heavy price. Police discovered him in the parking lot lighting the fire, and they took advantage of the situation and executed him on the spot, similar to how federal marshals would execute Michael Reinoehl in September, 2020 in retaliation to Reinoehl killing a white supremacist who was threatening Black counterprotestors at a rally in Portland several days earlier.[131]

Though the border regime is designed to brutalize racialized people and uphold the privileges of whiteness, these killings show that whiteness is less a question of skin color and more a question of alliances. Those who fight against the alliances represented by white supremacy and colonialism may face the very apparatus of annihilation designed to control racialized people.

Another major target of repression in the Global North are Indigenous communities, especially those resisting the profitable processes of ecocide that the reservation system was designed to enable. When Water Protectors tried to prevent the construction of the Dakota Access Pipeline transporting shale oil through Lakota territory, local, state, and federal police mobilized a huge militarized force complemented by the private security firm TigerSwan, complete with helicopters, drones, tanks, infiltration, and surveillance, constituting "the largest mobilization of cops and military in the state's history since 1890, when ... the military was deployed to crush the horseless and starving Ghost Dancers at Standing Rock."[132] TigerSwan agents, coming out of counterinsurgency operations in Iraq and Afghanistan, called Indigenous Water Protectors "terrorists" engaging in a "jihadist insurgency."[133] Private and public police forces carried out brutal raids, beat, shot, and gassed Water Protectors, and soaked them with powerful water cannons in subzero temperatures miles from any facility where they might receive adequate care for hypothermia. They used attack dogs, rubber bullets, concussion grenades, tear gas, and sound cannons. No one was killed in any of the multiple acts of police repression only thanks to good luck and well-organized medical support. However, a Native woman lost her sight after being shot in the face with a tear gas grenade at close range and a solidarity protester had her arm partially blown off by a concussion grenade.[134] Many others received traumatic head wounds from police projectiles. In one single attack on November 20, 2016, two hundred people were injured.[135] Multiple people were imprisoned with the aid of police infiltrators. Throughout the resistance, over eight hundred and thirty people were arrested and given state charges, while five people—all of them Indigenous—were given federal charges. One Indigenous and Chicano protester, Steve Martínez, is locked up at the time of this writing for refusing to testify before a grand jury.[136] He could be kept in jail for up to 18 months to force him to give information against other people in the movement.

It has been pointed out time and again that when far Right groups carry out armed occupations, even when they shoot and murder political opponents in the course of white riots, the police stand back and let it happen. Often, they participate directly. Subsequently, prosecutors are always more lax in bringing charges. This is not, however, a plea for fairness from the judicial system, or for more arrests of fascists. Rather, it is evidence that the fascists in the street, extractivist profiteers in corporate board rooms, and the criminal justice system are all facets of a white supremacist, colonial system that is destroying the planet.

Because repression transcends formal exercises of state power, we should also consider the connection between pipeline construction, shale oil extraction, and Man Camps, the itinerant labor camps of often highly paid, mostly white male workers associated with increasing rates of drug and alcohol usage and violence against Indigenous women and Two Spirit people. Going back to mining incursions in the Black Hills gold rush at the end of the nineteenth century or the US Army Corps of Engineers damming rivers in Lakota territory in the middle of the twentieth, settler profiteering has always been connected to violence against Indigenous communities, whether or not the perpetrators wear a uniform.

What happened at Standing Rock is not exceptional. It is fully in line with the US government's repressive practices against Indigenous resistance. In the early '70s, the American Indian Movement aided traditionals on the Pine Ridge Oglala Lakota reservation, not far from Standing Rock, to regain their sovereignty and expose the corrupt tribal government that was closely connected to Washington and to regional business interests, for example turning reservation grazing lands over to white ranchers. The BIA and FBI helped the tribal government stay in power, in part by financing a heavily armed paramilitary force. In the years that culminated in the 1973 Wounded Knee stand-off, FBI agents, police, paramilitaries, and unknown assailants killed over 60 Oglala traditionals, AIM members, and sympathizers.[137] Long-term prisoner Leonard Peltier is still locked up as a result of that repression.[138]

In Canada, progressive prime minister Justin Trudeau came into office on pledges of taking climate action and improving relations with the dozens of Indigenous nations whose stolen lands the Canadian state is built on. In fact, Trudeau took advantage of his conservative predecessor having left the Kyoto accords to go all in for the exploitation of the Athabasca tar sands, a huge deposit of extremely dirty oil that takes an enormous amount of energy to extract and process, leaving an immense expanse of the Canadian interior denuded of all forests and topsoil in the process. To capitalize on the oil boom, Trudeau's government sank billions of dollars into new pipeline construction despite fierce Indigenous resistance across the continent. When the Wet'suwet'en nation effectively blocked the construction of the Coastal GasLink pipeline on their own lands, the RCMP, Canada's federal police created initially as a paramilitary force against First Nations, carried out several militarized raids on Wet'suwet'en encampments and were given permission to shoot to kill. In their raids, the RCMP used boats, helicopters, drones, and intensive physical and social media surveillance of land defenders, in addition to large numbers of agents with armored vehicles and military grade weaponry. They also considered the use of social services to take custody of any children arrested, fully in line with Canada's long and violent history of using so-called residential schools targeting Indigenous children as a tool for genocide.[139] Canada's 2015 Anti-Terrorism Act "sanctions the criminalization of Indigenous environmentalists by enhancing surveillance and legal powers against any potential interference with Canada's 'critical infrastructure' or 'territorial integrity'" while RCMP surveillance documents identify land defenders as "aboriginal extremists."[140]

Again, recent Canadian history shows this is far from exceptional. In 1995, twenty First Nations recovered a part of the unceded Secwepemc territory at Ts'Peten, an important Sun Dance site. In response, the Canadian government sent in 400 police and military with helicopters and armored personnel carriers who fired 77,000 rounds at the warriors.[141] These are just a few examples. It would take more than a single book to document all the incidents of police and military repression against Indigenous peoples in Australia, Canada, the United States, and other settler states.

Powerful governments not aligned with NATO are no better when it comes to repression. In Russia, anarchists met with some successes protecting the Khimki Forest from highway construction in 2010. In response, police arrested anyone they could find associated with the protests and looked the other way as armed thugs attacked encampments set up in the forest to halt logging. As repression mounted, dozens of anarchists and antifascists fled the country, seeking asylum in Europe. The government turned towards preemptive repression. In 2017 and 2018, in advance of contested elections and the FIFA World Cup, nine anarchists were arrested in the "Network" case and sentenced to between six and eighteen years in prison. They were tortured into signing confessions, including being hung upside down and beaten, or being shocked with electric currents until their teeth broke.[142] Western media that usually enjoy publicizing Russian human rights abuses were largely silent on the case.

China has been the site of a growing number of "mass incidents" or riots, many of them against development projects, pollution, unsafe working conditions, and corruption between Party bosses and the new industrialists. A staggering 65 percent of the 180,000 annual "mass incidents" are rural conflicts triggered by land grabs, with government officials and private developers forcibly requisitioning village lands, sometimes without even offering compensation.[143]

Communication between these different sites of resistance has become increasingly difficult over the last ten years as the Chinese government consolidates its control over the internet. Signs of the underlying tension still boil over despite the censorship, as when a dispossessed villager killed four with a homemade bomb at a government office near Guangzhou this past March, amidst a conflict with farmers being dispossessed with little or no compensation for a major development project.[144]

The changing balance of power also means the overturning of earlier environmental victories, such as the defeat of a planned hydroelectric dam on the Yangtze River at Tiger Leaping Gorge. The dam, which would have displaced 100,000 people, was halted by locals in 2006, but now the government is resurrecting plans for the dam, reflecting a new balance of power, as well as the opportunities that "green energy" and "carbon neutral" policies afford governments for neutralizing environmental movements.[145] In another example of the same dynamic, three conservationists in the Ningxia Hui Autonomous Region were arrested and accused of extortion, making threats and "provoking trouble" in September 2020 after filing whistleblower complaints regarding pollution from a paper mill and destruction of protected gazelle habitat for the construction of wind farms.[146]

Both China and Cuba helped the US arrest Green Scare fugitives who were on the run. Furthermore, Chinese companies are fully invested in the fossil fuel industry and the destruction of Indigenous lands across the world, from tar sands extraction in Alberta to iron mining at sacred sites in Australia.

Green Scare prisoner Joseph Dibee, currently awaiting trial, wrote:

> The cataclysm that is unfolding today can be laid at the doorstep of the law enforcement agencies that have paved the way for it by making it so difficult for ordinary people to defend themselves against ecological devastation. If we don't stop them, they will frogmarch us directly into the apocalypse, profiting all the way—and when the last well is poisoned and the last forest burns up, they will be the last to die.[147]

A focus on law enforcement and repression more broadly enables us to understand the ecological crisis in a new light. We don't need the rich and powerful to save us from the destruction of the planet. In fact, we are the ones who are saving the planet every day. What we really need is for them to get the hell out of the way. But they won't do that, in fact they will systematically silence, marginalize, imprison, brutalize, torture, and kill those who are most affected by the ecological crisis, those who have the most collective experience and the best ideas for how to save the planet.

In this light, there is little difference between a coal mine and a wind farm or a hydroelectric dam. The affinity of green capitalism for the very same death squads, torture techniques, antiterrorism operations, and smear campaigns should belie the pretensions that so-called renewable energy or sustainable commerce is an improvement. The involuntary, authoritarian character of fossil fuel capitalism and green capitalism is impossible to ignore, once you start looking.

In December 2011, global private security firm Stratfor was hacked. Stratfor maintains high level relations with intelligence, political, and military officials from the US to Israel and provides consulting and intelligence services to private companies and governments around the world to aid in public relations, counterinsurgency, disruption, and regime change efforts. They have described themselves as a "shadow CIA."[148] The resulting data dumps revealed widespread surveillance and illegal actions by government and private entities. We only have access to this information thanks to the hack carried out by Anonymous. The anarchist hacker Jeremy Hammond was sentenced to ten years in prison for the action. Similarly, we only know about the FBI's bloody COINTELPRO program—which resulted in the aforementioned lethal campaign against AIM as well as the assassination of Fred Hampton and several other Black Panthers—because a group of revolutionaries broke into an FBI field office and stole the documents. In other words, the only way we ever know what our governments are actually doing is thanks to illegal direct action, leaving any discussion of government accountability rather hollow.

In the Statfor leaks, we learned how governments view social movements, and how they try to neutralize us. What is important to emphasize is that Stratfor, like governments all over the world, use the lens of counterinsurgency when dealing with the dissent of their subjects. With one exception, they do not make a distinction between the middle-class high school student going on strike on Fridays to raise awareness about climate change, the journalist researching the cancerous effects of widely used pesticides, the anarchist setting fire to construction equipment to save a forest from being bulldozed, and a Native elder leading a ceremony in an encampment seeking to regain treaty lands from a settler state. They view all these people as enemies. The only distinction they make is how to turn different kinds of dissidents against one another in order to buy out some and marginalize the rest. Remember that at the heart of counterinsurgency is the belief that all of a government's subjects are potential enemies.

According to a framework used by Stratfor and other security consultants, there are four types of dissidents and a three-step method to neutralizing them. There are the radicals, who "want to change the system; have underlying socio/political motives; and see multinational corporations as 'inherently evil.'"[149]

> These organizations do not trust the … federal, state and local governments to protect them and to safeguard the environment. They believe, rather, that individuals and local groups should have direct power over industry … I would categorize their principal aims … as social justice and political empowerment.[150]

There are the idealists who "want a perfect world" but because of their "intrinsic altruism" can be made to sympathize with the interests of industry and shy away from conflictive positions, there are opportunists who will seize "the opportunity to side with the powerful for career gain" and there are the realists who are "willing to work within the system" and are "pragmatic. The realists should always receive the highest priority in any strategy dealing with a public policy issue."

The three steps are rather simple:

> First, isolate the radicals. Second, "cultivate" the idealists and "educate" them into becoming realists. And finally, co-opt the realists

> into agreeing with industry. "If your industry can successfully bring about these relationships, the credibility of the radicals will be lost and opportunists can be counted on to share in the final policy solution."[151]

It is more than a little ironic that those who consider themselves realists are considered the easiest to manipulate.

This is why it is neither sectarianism nor an excess of zeal when we declare that NGOs and humanitarian activists are part of the state's counterinsurgency campaign. We can see this when we look back at resistance to the oil industry in Nigeria. When Ken Saro-Wiwa was led up to a scaffold and executed, his death was counted. He was a poet and a nonviolent protest leader, after all.

But what happened to his peers who learned the lesson, who valued their own lives, who took up arms or gave shelter to those who did, and *who actually hurt the profits and cut off the production of the oil industry*? When they were murdered by British or US munitions fired from helicopters dispatched by the Nigerian government or Shell Oil, those very well-meaning NGOs did not include them in their list of murdered land defenders. Their lives are neither counted nor valued.

The road to hell is paved with good intentions. To those with blood on their hands, counterinsurgency revolves around separating the resistance into good actors and bad actors, negotiating with or buying off the good actors so that the bad actors can be isolated, imprisoned, or massacred. Not so many people would trust Stratfor, Shell Oil, or the US military's Africa Command telling us that Ijaw and Ogoni people fighting for their homelands are evil terrorists who deserve to be shot down. It is the very NGOs who evince a concern for human rights that are crucial to this counterinsurgency operation, celebrating the activists who use tactics deemed legitimate by the government. By not celebrating every bit as vociferously the lives and the resistance of those who make the hard choice to resist with illegal and revolutionary methods, these NGOs are signaling who is legitimate and who is subhuman, they are creating the division between good actors and bad actors that the militaries, death squads, and counterinsurgency experts rely on, and they are doing so in a way that most people who think they care about the environment or human rights will find credible. Likewise, they are signaling to the rest of society what kinds of tactics (legal, peaceful, less effective) will be honored, and what kinds of tactics (illegal, transformative) are associated with dangerous, frightening elements who seem to inevitably—in the moral arc of capitalism's universe—end up in unmarked graves.

These NGOs also reject the learning process that happens in communities that, through direct experience, turn from methods that revolve around petitioning, protest, and lawsuit, to methods that revolve around occupation, sabotage, and directly creating the changes they need. As long as social movements have memory, they have access to a vast store of collective intelligence, and this intelligence favors self-defense and direct action because such methods are more effective.[152] By rejecting such methods, the NGOs place themselves above the communities in struggle, claiming to know best how to achieve change. This is nothing other than a continuation of colonial dynamics.

In order to outmaneuver these counterinsurgency strategies, we need to start by becoming aware of them. There can be no effective struggle for healthy lives on a healthy planet unless we are aware of the systematic repression leveled against those who seek this most basic goal, and stand in solidarity with those who are most affected. Because when we support one another and fight back, we can win.

3
The Solutions Are Already Here

WE HAVE STOPPED PIPELINES, AIRPORTS, HIGHWAYS, AND MINES: THE VICTORIES THAT ADD UP

Whereas governmental and market responses to the ecological crisis have an almost perfect record of failure, there is in fact another methodology that has achieved stunning victories and concrete, real gains across the world. After the previous chapter, it should come as no surprise that this methodology is almost completely invisibilized or dismissed as a series of isolated acts of resistance that are interesting anecdotally but that could never "scale up" enough to become significant.

On the contrary, the network of resistance we are about to dive into has access to greater local knowledge than any government or scientific institution; rather than being isolated we in fact network and debate strategies globally, confounding the borders, the unequal access to resources, and even the internet firewalls erected to divide us; and several times we have demonstrated the ability to scale up our actions across an entire continent far more rapidly than capitalist and state bureaucracies are capable of. Rather than using technocratic blueprints and simplified statistics as our starting point, comfortable with the idea of imposing our solutions on populations we deem as unqualified to take part in designing them because they do not hold the right degrees, our starting point is the desire to protect our territories and our livelihoods, both of which we understand intimately, far better than any expert.

And the fact of the matter is, if we win, capitalism becomes impossible, and with it, ecocide on a global scale. While these movements need to improve their global visions and their plans for a safe transition away from ecocidal society—and we are in fact in the process of doing so—time and again we have demonstrated how quickly we can adapt to incorporate new techniques and changing circumstances.

The struggles that follow have achieved concrete gains in stopping the infrastructures and the extractivism that is killing the planet, changing the conditions on the ground that make ecocide possible, rather than simply shuffling the problem around through carbon trading, future emissions promises, or green industries that carry out ecocide in slightly different ways. The main obstacle to the models presented here scaling up, stopping climate change, and healing the planet are government repression, police and paramilitary violence, and the media, NGOs, and academic institutions that either naturalize repression or turn a blind eye.

Dayak Tomun and Dayak Tamambaloh, Indonesia

Indonesia is home to the third largest tropical forest in the world, an important site of biodiversity and a vital carbon sink for the global climate, but it is now being deforested faster than the Amazon. One of the main drivers of the destruction, responsible for 40 percent, is the palm oil industry, which seizes and clears land for destructive plantations.[1] Indigenous communities and small farmers are often violently evicted, and local companies clear the forests with aggressive burning. In 2019 alone, 10 million children were put at risk of air pollution from the fires, which released 360 million metric tons of carbon dioxide in just a month and a half. Palm oil fires in 2015 caused an estimated 100,000 premature deaths.[2] Those responsible for the plantations and processing operations include Indonesian, Malaysian, Singaporean, and US companies, while the major buyers are global giants Unilever, Nestlé, and P&G, corporations with deeply colonial histories based in the UK and Netherlands, Switzerland, and the United States, respectively. The largest importers of Indonesian palm oil by country are India and China, while the logging industry—alongside palm oil one of the major causes of deforestation—exports primarily to China and Japan. Only about a quarter of Indonesian palm oil is consumed locally, as cooking oil.[3] Aside from its extensive use in processed foods and cosmetics, the growth in the palm oil industry is being fueled by its use as a supposedly clean biofuel.

I spoke with an anarchist activist and researcher from a mixed Dayak family from Kalimantan (Borneo). They asked to use a pseudonym for the interview to avoid police attention:

> You can call me Jungkir Maruta. I was born and raised in Borneo, and since I was little, I had concerns about environmental and agrarian issues where I live.
>
> Many Indigenous communities in Kalimantan are fighting against palm oil plantations. One of them that received national attention is

Dayak Tomun at Laman Kinipan, Lamandau, Central Kalimantan. They are against the expansion of a palm plantation owned by PT SML. But now I spend more time with Dayak Tamambaloh in Kapuas Hulu, West Kalimantan, who have also taken preventive action by blocking the clearing of land for palm oil for several years and successfully defending their forests (although in this case, they are dealing with a National Park denying their access to forest products that they have been using for hundreds of years).

These two Dayak tribes have a very high level of dependence on the forest, also because their village is so far from the market. They are shifting cultivators who require a large amount of land. They can meet their basic food needs by producing their own rice. Apart from that, the need for food, housing, and making household utensils is still dependent on rattan and other forest products. They also still set traps to catch deer and pigs, and they catch fish. They grow vegetables mixed with wild food and one of the tribes still uses traditional medicine and traditional healers. Even so, the tendency is for this knowledge and dependence to decrease as a result of the penetration of capitalism and the state. What makes them safe enough to this day is only that they are an isolated community. I don't know how long it will last.

Like I mentioned, the Dayak Tomum are fighting the expansion of an oil palm plantation owned by PT SML. The company executives are also connected to local officials, like Sugianto Sabran, governor of Central Kalimantan [linked to the violent attack on a journalist mentioned in Chapter 2]. The local media called him the "ambassador of the oligarchy." He is the nephew of Abdul Rasyid and H. M. Ruslan (both members of the Golongan Karya Party, Golkar), the owners of PT SML. Rasyid, the boss of the Tanjung Lingga Group and a former MPR [parliamentary] representative for Central Kalimantan, together with his younger brother, Ruslan, started an illegal logging business during the New Order era. He was listed as one of the 50 richest Indonesian people according to Forbes Magazine for two consecutive years. Meanwhile, Ruslan's wife, Nurhidayah (of the Golkar party), is the current Regent of West Kotawaringin and was appointed by her nephew, Sugianto. Several other public officials in Central Kalimantan also come from their families.

When the forest was depleted, in 1995 Rasyid established PT Sawit Sumber Mas Sarana (SSMS) creating oil palm plantations on 100,000 hectares (in 2016) of state-owned land and making a profit of IDR 787.1 billion in the first quarter of 2018. Illegal logging and persecution by this oligarchic dynasty is recorded in dozens of white papers and many environmental agency documentaries. Sugianto was suspected of being involved in the pillaging of a forest in Tanjung Puting National Park when he was entrusted with managing his uncle's company that owned a plantation of more than 40,000 hectares, and he hired thugs as company security guards.

When asked about government support for the companies responsible for deforestation, I have to emphasize that entrepreneurs are the government in Kalimantan (and in fact throughout Indonesia).

Despite having the entire power apparatus arrayed against them, Dayak communities have put a stop to land clearing in Lamandau. They took direct action by seizing heavy equipment, evicting palm oil company employees, clashing with police, and blockading roads. But I think it's only a matter of time before it resumes. Not many struggles in Kalimantan have been successful, and land grabbing is so common.

The Indonesian state is very repressive. Visually, black bloc anarchists dominate the images in the national media, so the government has new enemies to blame. This is getting easier, because they associate anarcho-syndicalism as a new variant of communism.

In Kalimantan, Indigenous communities are at the forefront of the struggle, with anarchists slowly setting up campaigns of solidarity. There are also some mass organizations that exhibit ethno-nationalist sentiments. I fear that if alternative discourses are not rolled out, they will turn out to be worse than what they are against.

Sápmi, Northern Europe

The Sámi are an Indigenous people within the European subcontinent. Their territory, Sápmi, is largely within the Arctic Circle and occupied by the states of Norway, Sweden, Finland, and Russia. Many Sámi are engaged in fights against industrial wind farms and mines that poison the water and destroy lands needed for traditional reindeer herding. Much of their land has also been taken over for military training exercises and bases, as well as tourism. Blockade camps and protests have stopped iron mining in Gállok, at least for the moment, though "the Swedish government has not yet decided how to proceed."[4]

Aslak Holmberg, "a salmon fisher, Sámi language teacher, indigenous studies scholar, and Vice President of the Sámi Council"[5] spoke with Gabriel Kuhn about resistance to new fishing regulations passed in 2017 by the Finnish and Norwegian governments for the Deatnu River, which marks the border of the two states for over two hundred kilometers and is part of occupied Sápmi.

> Sámi live on both sides of the river, and they've been fishing salmon in it for centuries. The new regulations affect primarily our traditional ways of fishing.
>
> ... The Deatnu River is one of the very few salmon rivers where you can fish with nets. That's what we do. But rod fishing has higher status. If you look at the new regulations, there is only one very specific group of fishers that gains something: the Finnish cabin owners who have bought property along the river with fishing rights attached to it. ...
>
> [The new regulations] are supposed to protect the salmon. But essentially they are just shifting fishing rights from one group, the Sámi, to another, the cabin owners. The consequences go far beyond fishing. Our entire culture is based on the fishing tradition, so when our fishing rights are restricted, our entire culture is under threat.

In 2017, local Sámi people began establishing a camp on an island in the Deatnu River every summer to enforce their moratorium on the new regulations. They would only allow fishing in accordance with traditional practices, and tourist fishers would have to specifically ask for permission from the Sámi of the Deatnu valley, respecting the traditional areas of local families.[6]

Niillas Somby is a veteran of the Sámi struggle going back to the 1970s. After being severely injured during the Áltá movement when a bomb went off prematurely during an action against a power plant in 1982, he evaded the authorities "and subsequently found refuge among First Nations in British Columbia. He has worked as a reindeer herder, sailor, mechanic, photographer, and journalist."[7] He is also a descendant of one of the people executed by Norway in the Guovdageaidnu Rebellion, and he fought to have the skulls of the executed returned by the state for proper burial.

He speaks about some of the dangers of success, and how the victories of past movements have created channels of institutionalization that can weaken their struggles today.

It was great that the Áltá movement put Sámi rights on the political agenda. But it also established a new class of Sámi politicians who soon claimed control over how these rights were to be implemented. Sámi activism was integrated into the colonial system, and people like us were told to be quiet.

... In some ways, things have become more complicated because of the Finnmark Act. Before it was passed, we could go to the forest and collect firewood. Now we have to pay for it. It feels like we have given away land with the stroke of a pen, rather than ensuring that it is ours...

The Sámi Parliament has consultation rights. That is as much as it has achieved. But what are consultation rights? It means the authorities are obliged to meet with you and listen to what you have to say—before doing whatever they want to do.[8]

ZAD, France

In 2018, a long-term land occupation at Notre-Dame-des-Landes, near Nantes in northern France, successfully stopped the construction of a major airport. When the government began planning and land seizures for the mega-project, they gave it the bland, bureaucratic name of the ZAD, from "*zone d'aménagement différé*" (or "zone to be managed"). The motley alliance of farmers and anticapitalists who refused to abandon the wetlands, forests, and farmland re-baptized it the "*Zone à défendre*" (or "zone to defend").[9]

Blocking the airport and saving the wetlands was a major victory, coming at a time when the European economy was booming, carried in large part by tourism and cross-border business networks that favored a dense infrastructure of short-haul flights. But the struggle was about much more than just an airport, and painful divides grew between participants who had differing visions of what they were fighting for, what kind of world they wanted to live in, and what were acceptable methods for getting there. Going beyond the immediate victory to understand how the struggle fell short is crucial as we go from fighting specific battles to unleashing a wave of global transformation. The accounts in this piece come from three long-time ZAD residents. I asked them not only about what they achieved, but also in what ways their struggle failed. For reasons of state repression, they preferred to remain anonymous.

When we speak from our own perspective in the struggle against the airport (and its world) we speak in the past tense because for us the struggle on the ZAD is over, even if people still live there. There have been many different phases in this struggle, because it lasted for such a long time, so it's hard to speak in absolutes.

It was a struggle against an airport, against the world that needs such airports, against the state and the state's ability to decide how our lives and the places we live are structured, and it was a struggle for community self-determination. There were many different kinds of people who lived there and who participated in the movement against the airport: urban squatters, environmentalists, small farmers, and street punks.

"The ZAD occupation" and "the movement against the airport" were not the same thing. Over time there was a divergence between "the ZAD"—which was sometimes more concerned with daily life, solidarity with each other, and material and social collective structures—and "the movement," which drew from a larger geographical area, had many different participating organizations, and was more liberal. The general assemblies of the movement increasingly became the terrain of "little p" politicians (from both on and off the zone) while the voices of many other people and groups who lived on the ZAD became increasingly pushed to the margins or ignored. The committee that tried to control the assemblies derived their power from "the movement," whereas the ZAD occupation was messy, diverse, and principled and got in their way.

When the first plans for the airport were announced in 1963, locals organized together, some against the airport and others to collectively negotiate eminent domain and avoid getting screwed over for the value of their farms. The project was dropped in 1972 and restarted again in 2000. With the new version of the project came a new version of local resistance: ACIPA, an anti-airport association started by local farmers and residents. In 2009, a number of local residents facing expropriation formed a group called "the resisting residents." They made a call-out for squatters to come and occupy the empty houses on the zone that had already been seized. At the end of the 2009 Climate Camp, which was held on the ZAD, ten people decided to stay and occupy houses, beginning the strategy of occupation as a means to fight the project, and joining a small group that had been squatting a house on the zone since 2007.

The effectiveness of the ZAD was not a magical accident: it was built on a history of environmental, social, and agricultural struggles. A culture of resistance permeated the territory, with lots of local people participating. The small farmer and worker movement that was strong in this area in the '70s was one of the things that set the stage for this struggle, when farmers and other workers found class solidarity together during strikes. The small-scale farmer defense organization "SOS Paysans" had existed for decades, and there had been prior victories of small farmers fighting for access to the land. On a larger scale, there were two well-known environmental and anti-state conflicts in the '70s—in Larzac and Plogoff—against "*aménagement du territoire*" which roughly translates to "technocratic development planning."

At the beginning of the ZAD occupation, the people involved were mainly from the French urban squatting movement, influenced by feminist and anti-authoritarian politics. Many people had been radicalized in the CPE social movements of 2006, opposing a major reform that would increase job insecurity for young workers. They were joined on the ZAD a few months later by people from the network of "anti-national" forest occupations across Europe.

The struggle succeeded in preventing the airport from being built. It created political bases for collective ownership and management of the land, particularly agricultural land. It allowed people to live together in an autonomous, self-organized way; creating social and material infrastructure to survive outside of state and monetary systems, instead relying on sharing and solidarity. It allowed many people in France and beyond to see that it's possible to struggle, to resist, to organize—concretely, in practice. In this way it changed the perception of what was possible for the generation that was growing up at the time. There are adults on the ZAD who talk about arriving in 2012 during the first round of evictions as pre-teens with their parents and how that affected them. The ZAD also managed to create an incredibly powerful political *rapport de force* out of basically stubbornness and conviction. In some cases, the ZAD was able to make workers involved in the development and state apparatuses realize what their jobs entailed, and encourage them to quit (especially in the case of environmental consultants, but even cops too). The occupation created an internal conflict resolution structure that lasted for several years to manage interpersonal disputes in the absence of police and courts. There was also the creation of a horizontal and diversified healthcare system that ensured people had access to the care they needed, regardless of citizenship, papers, or wealth. It also proved to be skilled and effective at dealing with the consequences of police violence on the ground.

The ZAD also fostered the presence of radical politicized thought, connecting different ways of engaging in daily life struggle through participation in a diverse political community. It's important to mention the contributions made by those who are normally marginalized in society, such as people living with addictions, street punks, homeless people, and migrants. It was both an asset and a liability for the ZAD that it was a zone open to anybody—all kinds of people from different walks of life came to visit or to live there. Anyone could come, anyone could stay.

Furthermore, the ZAD became a political model for fighting infrastructural mega-projects imposed on the population, and that idea spread across France. More than 20 land occupations that called themselves "ZADs" have existed between 2014 and the present. While the land was squatted (until 2018), we often heard the ZAD described as an "open air laboratory of alternatives." Hundreds of people organized without the state on a large area (8 square miles) of countryside and experimented with all kinds of different forms of autonomous organization on a mid-size scale, confronting theory with reality.

A key player in the movement against the airport was COPAIN, created to advocate for small-scale farming over industrial agriculture. The group was made up of farmers who were fighting to have their own relationship to the land in their farming practices, and not be directed by engineers and machines. There was also the creation of "*Sème ta Zad*" ("Sow your Zad") or STZ, in 2013, a collective structure for organizing between everyone who wanted to use agricultural land on the ZAD, including members of COPAIN, squatters, and other farmers or people with agricultural projects. Together, STZ decided crop rotations, organized collective money for seeds, tools, and maintenance, and organized regular upkeep and repair of tractors and other collective machines. The movement defended respectful agricultural models, on a basis of a shared commons rather than private acquisition.

There were deep divisions about the relationship to the land on the ZAD. From early on, there were people who wanted to let nature do its thing, who advocated doing everything by hand, and who were angry about pesticide use by local farmers. For example, a neighborhood in the east of the zone closed its paths to motor vehicles using barricades from 2013 to 2018. On these 66 acres there was no electricity or running water. This neighborhood championed a "non-management" of land, with the goal of letting part of the occupied land return to a more wild state, with some cabins and small permaculture gardens that blended in with the landscape, including walls made from local clay and living trees. People there often foraged for food and healthcare. This relationship to the world was central to discussions in this neighborhood, in what they created and imagined together, in collective activities and in personal reflections. A deep ideological conflict regarding humans' relationship to nature situated the squatters of this neighborhood in opposition to local farmers in the anti-airport movement and to other groups of squatters. The state clearly took advantage of this dissension, targeting the groups of squatters it saw as most resistant to assimilation. All living structures and gardens in this neighborhood were entirely destroyed in the 2018 evictions.

The relationship with the state was conflictual from the start. We were fighting against capitalism in the form of private property, monetary exchange, systems that put monetary value on living things and ecosystems thus profiting from their destruction, and the logic of "environmental compensation." We were fighting against all the companies that collaborated with the state to promote the project—the contractor VINCI, subsidiaries like Eurovia, as well as subcontractors like Biotope who performed environmental impact assessments ("quantifying" the wetland ecosystem so its destruction could be "offset" by digging ponds somewhere else).

As the Prefect of Loire Atlantique said in October 2012, "If the republic can't take control of the Zone, then we need to be worried about the state of the republic."

The state tried to crush us with eviction and military force in October 2012. There was fierce resistance and a lot of media coverage that led to the ZAD becoming a household name and gaining massive support. Not only did the state not manage to destroy all the houses and take back control of the territory, the attempt made them look both overly repressive and ineffective. It was very embarrassing for them and they effectively lost control of the zone for several years afterwards. Leading up to the evictions in 2018 they were much more strategic—they wanted to be sure they'd win no matter what. The state abandoned the airport project because it became too difficult for them, and also to neutralize the ZAD's political power. With the relative loss in mainstream support once the "cause" was "won," it was much easier for them to proceed with evictions.

The state tried to break the mutual support that gave us strength, to divide the local population and the wider movement in many ways—continuing with land expropriations to threaten and physically remove local resistance, having the regional population vote in a referendum, and setting up a series of negotiations with representatives from different groups in the movement. The negotiations took place with representatives from several groups—ACIPA, a local farmers organization, Naturalists in Struggle, the farmers' organization COPAIN, a coalition of organizations against the airport, and squatters from the ZAD. These negotiations ended up being one of the most successful parts of the state strategy for sowing divisions. Many counterinsurgency tactics were used to create fractures in the movement and turn the local population against the ZAD. The media fully participated in this.

Our interactions with the police changed a lot over the years. At the beginning of the occupation the police were very present but pretty ineffective. They harassed us but didn't do much else, and as we generally refused to identify ourselves, there wasn't much they could do. It's hard to issue a summons for "X," and it was annoying for them to waste their time and paperwork knowing we'd just refuse to identify ourselves. Several times they even suggested we just give a fake name to be able to get rid of us. This made us all a lot more daring and confident. The police presence and other forms of state violence started to really ramp up after a while, in demonstrations and especially in the 2012 evictions. The 2018 evictions were incredibly violent, even compared to "normal" French police repression of urban riots. It is truly incredible that no one was killed at that time, as happened in the eviction of another ZAD in Testet in 2014.

After the 2012 evictions there were 6 months of military occupation, all the roads leading into the ZAD had police roadblocks 24/7 with large spotlights. There was a big confrontation in April 2013 with one cop set on fire and the police left afterwards. They were not allowed to come onto the zone for the next 5 years because it was "too dangerous." In our efforts to prove the viability of living free of police and the judicial system, we organized horizontal structures, such as the "Cycle de 12" to deal with interpersonal conflicts.

During the years without the police on the ground, there was still regular surveillance by helicopter. The vast resources deployed by the state demonstrated their fear of another failed operation. In advance of the 2018 evictions they enlisted a General to lead preparations, using drones, tanks, an intelligence-gathering and listening van, and visible assault weapons. We have since learned that the military police had integrated a "Notre-Dame-des-Landes" exercise into their training programs for both cadets and officers, and that the *prefect* who oversaw the 2018 evictions now teaches classes about this period in famous schools of political science. Furthermore, the French state now sells to other countries the counterinsurgency tactics they trialed on us.

We protected ourselves from the police by creating a culture of resistance—using pseudonyms, not revealing our identities when arrested, refusing photos, fingerprints, and DNA testing, communicating about actions with little printed paper invitations that had a time, place, and a risk level so that information was never spoken out loud (to foil surveillance via phones and microphones). There were incredibly strong bonds of trust between us, with collective physical resistance in case of arrest and collective defense in trials. The legal team did direct support for arrestees and prisoners, coupled with political work against repression. After the police left in 2013, there was suddenly no more daily confrontation with a common enemy, and we fought more among ourselves. While the legal team continued their work, we lost a collective transmission of those practices of refusal to new arrivals, as daily encounters with police became rare.

Up until the movement became more consolidated or centralized around 2016, it was an excellent argument for the effectiveness of using a diversity of tactics. It was impossible for the state to counter a mix of nonviolent liberal demonstrations, court challenges, university educated squatters of multiple persuasions, and completely unpredictable wild-card street punks at the same time. The more the movement pushed for unity around an unconsensual strategy and enforced adherence to this top-down unity by dissuading actions deemed to threaten it, the weaker the whole struggle became. Some groups' motivation to negotiate with the state on their own terms required them to flatten and shape the resistance into something that the state could understand. Eventually, even sabotage was violently discouraged, because it might hurt the negotiations. Unity was a strategy that failed—as enforcing this idea of unity around a common project involved the exclusion of dissenters, the links that had previously held us together were broken.

Up until that point, a wide variety of tactics and strategies had been used. This included everything from festively costumed mass bike rides through one of the nearby cities while carrying out direct actions or attacks against businesses and other targets related to airport construction, marches with agricultural implements that ended in collective land occupation, squatting land for farms and orchards, *autoreduction* (mass refusal to pay at the supermarket), mass demonstrations, raiding offices for inside information, barricades, catapults, infotours, writing songs for the movement, house-squatting theater pieces, eating together and providing food for conferences and protests, ambushing the police, taking care of local people's cows while they were on hunger strike, sharing activities and daily responsibilities that let us create and strengthen relationships within the occupation and the movement. There were acts of sabotage, blockades, media actions, legal appeals, collective decision making, general assemblies, and organizing spread throughout France with "local support committees."

The mainstream media only repeated the discourse of the state at first (partly because we refused to talk to them). They gave a loudspeaker to all kinds of politicians who saw in us "the great danger," an outlaw zone that needed to be flattened to save the republic, a dangerous domestic enemy. *Le Journal du Dimanche* wrote an article in 2018 based on "secret" photos, which were just everyday photos of the ZAD, described in fearmongering ways—for example, a hand dug well was labeled a "weapons cache." We can also cite the former president of the region, Bruno Retailleau, in 2017: "Notre-Dame-des-Landes has become a symbol. A political symbol. And Emmanuel Macron should be on his guard because what is at stake is nothing more and nothing less than to know if an ultra-violent minority will manage to push back the rule of law … If Macron backs down, he will send to all French people the terrible signal of the collapse of the state."

The more well known the struggle became, the more the media had a role to play—in stigmatizing and dividing the participants, trying to sort and label the squatters and create a difference between the good (young, educated, productive, middle-class, white, farmers) and the bad (violent, black-bloc, drug-addicts, window-breakers). At some point a group of people arrived on the zone who played right into the media game of "good vs. bad squatters." Some people couldn't help but project a good image and be loved for their adherence to social and class norms. For a long time we had collectively chosen anonymity and had a distant relationship to the media—we considered that they needed us more than we needed them. Around 2016 when the press group was reinstated, there was an established norm where the journalists would only film hands and feet instead of faces, unless people were masked. This prevented us from having an identifiable spokesperson, and also made it possible for anyone to talk to the media, even if they were wanted by the police. Some people in the new version of the press group disliked this approach, saying it made them uncomfortable and "looked weird," and started giving interviews unmasked, making it much harder to get the press to accept masked interviewees and so reducing the diversity of those speaking to the media.

As for autonomous media, there was a pirate radio, website, newspaper with position essays and news, communiques, and actions to spread ideas and perspectives.

While the relationship between the ZAD and journalists was mostly antagonistic to begin with, media opinion changed as the ZAD struggle continued and grew. Some journalists shifted their perspective and tried to understand the phenomenon. Prior to the 2018 evictions, the cops said journalists weren't allowed in, because "they (the police) would provide adequate footage." The journalists came anyways, and got shot at with grenades like everybody else, even though they were wearing high-vis jackets that said "press." Some of them were treated by ZAD medics, and the medic team and some journalists later filed a joint formal complaint together to the human rights ombudsman.

The horizontality of the general assemblies took several large hits over the years, with a few different groups and personalities monopolizing most of the space. There is a lot written about this elsewhere, notably "Reflections on the ZAD" and "The Movement Is Dead, Long Live Reform."[10] One group in particular took increasingly more power, through increasingly less subtle means. This committee, whose existence as an organized political group with a common strategy was unknown to most of those living on the occupation, seized power in numerous ways. One of the ways they consolidated their position was to court and emphasize links with the leaders of important groups in the movement.

In order to be taken seriously by those moderate groups, they needed to prove themselves capable of reinforcing control of the rest of the occupation—who were much less willing to sacrifice radical political goals and tactics in exchange for increased mainstream support. They thus distanced themselves from the existing internal political structures of the ZAD, delegitimized them, and created others that were both easier to control and more acceptable to other components of the anti-airport movement. They took control of the communication and press groups. When their machinations came to light and they were challenged by others on the ZAD, they refused to change their practices. The general assemblies became a tool to channel power and legitimize decisions that had already been made elsewhere, killing possibilities for horizontal organizing in the name of speed and efficiency. In the lead-up to and during the 2018 evictions, it was this same group who were outspoken critics of direct action against the police presence and who pushed the idea that negotiations with the state were inevitable. In the end, they got what they wanted and today are a large part of those who live legally on the land that used to be the ZAD.

When evictions were announced, several leaders of the off-zone components of the movement, who we had worked with for almost a decade, publicly separated themselves from the eastern part of the zone—a geographic area where a large part of the occupants were radically opposed to any form of negotiation with the state. These movement leaders stated they would only defend the Zone up to a certain geographical boundary, despite their claims of unity. Others met privately with the chief of police, and then gave interviews to the media on the eve of evictions—calling for the eviction of certain areas and demonizing squatters. These actions weakened public support, putting the lives of those resisting at risk by legitimizing the state narrative of "by any means necessary."

The currents that made effective resistance more difficult were all those who used the struggle to advance their personal or collective interests at the expense of the struggle, the wannabe politicians among us and the organizations that were anchored in social norms, respect for the state, and desire for a return to the status quo.

Some of the other ZADs that have been formed have succeeded in blocking infrastructure projects in other parts of France, and many have put their collective experience to use to overcome the authoritarian dynamics that prevented the occupation at Notre-Dame-des-Landes from expressing its full potential.

Protest camps have been an important method for stopping extractive industries throughout Europe. This phenomenon, as differentiated from rural communities defending their territory against extractivism, is likely a result of the countryside being largely depopulated in wealthy countries, with remaining populations already largely integrated in extractive industries. As such, protest camps and similar experiences are a vital way for alienated populations to reconnect with the land and break with a technocratic view of nature.

The anti-roads movement in the UK saved several remaining forests from a massive expansion of motorways the Thatcher government set in motion in 1989. Saved forests include Oxleas Wood in Greenwich and Bilston Glen in Scotland. Even the protest camps that were unsuccessful often doubled the costs of road construction and generated public opposition, such that in the latter half of the '90s, the government canceled over 400 projected road constructions, though some of these projects were put back on the table a decade later.[11] The movement was also an element in the creation of "Reclaim the Streets," an urban event focused on opposing car culture and reclaiming cities for their inhabitants through unpermitted, roving dance parties and temporary autonomous zones.[12] The practice, which spread to over a dozen countries on at least four continents, intuitively overcomes the separation between urban and rural, human and environmental issues, and mixes questions of culture, quality of life, survival, ecology, and confrontation in a way that still eludes NGO-oriented environmental movements.

Long-term tree sits taking place in a context of environmental protest and sabotage against the logging industry saved numerous forests on the west coast of the US in the 1990s and 2000s. And it is a tactic that is being used today in many other places.

Over the last ten years, forest camps and related resistance movements have proliferated across Germany. The threats to the forests have been manifold, all related to industrial capitalist practices: open pit coal mining, gravel mining, lime pits, factory expansions, waste disposal, and road construction. Responses to the forest camps have been fairly uniform: a combination of state repression, mercenary violence, and political cooptation through dialogue and false promises. At least one person has died in the course of forest defense and evictions, thousands have been injured by militarized police forces and in some cases by corporate mercenaries, and many thousands have been arrested and saddled with drawn out court cases, debilitating fines, and jail sentences.

While media and politicians demonize those who physically stop the devastation, preparing the field for harsher repression and even lethal force, they celebrate activists who enter into dialogue and claim to share the same priorities of protecting the Earth. We already know where that road leads. In several German states where they are part of the ruling coalition, the Green Party are the managers of the devastation, supporting profitable highway development schemes and the attendant destruction of significant forest land.

The forest defense movement has protected the Steinhauser Forest from the planned expansion of a factory, and after over a decade of struggle won the protection of a small part of Hambach Forest against a monstrous lignite coal mine owned by the major energy corporation RWE and vociferously supported at the highest levels of the German government, marking a strategic priority for state power. Though most of Hambach Forest has been destroyed, the movement to save it has inspired a proliferation of similar struggles and raised public awareness about a previously invisible problem, put pressure on the German government to begin phasing out coal, and set the stage for future victories on a greater scale. The resistance also constituted a learning process about the futility of legal protests, as well as a laboratory for effective methods of resistance.

The forest defense movement utilizes a diversity of tactics. Included in the blend are the student strikers of Fridays for the Future, the wave of mass civil disobedience called Ende Gelände, which explicitly avoided use of the term "nonviolence" to sidestep any self-defeating tactical handicaps. There are long-term tree villages and forest occupations, barricades, urban protests, and numerous actions of sabotage including the burning of vehicles, blocking of train tracks, and arsons of power lines to shut down a major coal mine and coal-fired power plants; these actions also cause significant economic losses to the companies responsible. A culture of rich internal debate accompanies this heterogeneity, for example on the question of whether to dialogue with politicians or reject them, how to combine different tactics, and how to communicate with a broader section of society and challenge the distortions spread by the media.

The movement also exhibits a close relationship between rural forest camps that are anti-industrial in character, and inner-city tree sits related to the concept of a *right to the city*, opposition to car culture, and horizontal direct action to improve urban health. This reciprocity across and against the dichotomy of urban and rural will be a key feature as these movements become stronger and more intelligent over the next decade.

An in-depth analysis of the strategy and history of the German forest defense movement, published anonymously by the anarchist group CrimethInc, points out the importance of local initiatives, often passed over because of their small-scale, in launching major movements. "Local initiatives from those who are directly impacted by the things they are protesting are a crucial element in the success of large movements. Local expertise and continuous work over years and decades can neither be provided by activist groups nor by NGOs focused on nationwide work."[13]

Indeed, the movement has expanded considerably, with forest defense movements in over a dozen locations across Germany, networked with forest occupations in Poland, Switzerland, France, and other countries. In one example, anarchist squatters and tenants' groups using tree sits and blockades protected the city's green belt from a development company in Poznań, Poland.[14] The green belt provides working class neighbors with allotment gardens and a healthier environment, whereas gentrification threatens the trees, the gardens, and the neighbors.

An even more powerful example of networked resistance, spreading in this case across an entire continent, can be found in the blockades, camps, and protests against fossil fuel infrastructure in territories occupied by Canada and the United States. The resistance is centered in Indigenous struggles to assert their autonomy and win their land back. These struggles have increased in strength since the Oka standoff in 1990, in which Mohawk warriors from the communities of Oka, Kanesatake, Kahnawake, and Akwesasne defeated an attempt—backed by the armed force of the Canadian state—to expand a golf course on their lands. Other struggles that built up this resistance and increased solidarity among Indigenous peoples include the 1995 attempt to recover a Sundance camp in Secwepmec territory (occupied by British Columbia), defense of traditional hunting and fishing rights by nations such as the Mi'kmaq, Ojibwe, Haudenosaunee, and Yakama, resistance by the Six Nations of the Grand River to suburban development on unceded territory near the Toronto metropolitan area, the long-running opposition to coal mining by the Diné at Black Mesa, near the Grand Canyon, and resistance by the Yaqui and Tohono O'odham to the militarization of the US-Mexico border, which runs through their territories.[15]

The fossil fuel industry has been a frequent target of resistance. In 2013, people of the Elsipogtog First Nation blockaded highways and seismic equipment, and lit a sacred fire to stop an energy company from fracking in the region. When police moved to evict the blockades, intense fighting broke out and six police cars were seized and set on fire. The resistance led the company to suspend their activities for two years, during which time the government of New Brunswick imposed a moratorium on fracking throughout the entire province to avoid more conflict.[16]

Wet'suwet'en opposition to the construction of pipelines on their land led to a new level of resistance. A boom in oil and gas exploitation—including the extremely ecocidal Athabasca tar sands mining—supported by the Canadian government and various multinational corporations, has required an expansion in transport infrastructure, including the Coastal GasLink (CGL) pipeline which is intended to bring gas from the interior to a port on the Pacific coast for shipping to Asian markets. In 2010, Wet'suwet'en hereditary chiefs and communities blocked an earlier pipeline project by setting up encampments on their lands, blocking the pipeline's path. In 2018, they continued this tactic against the CGL, leading to multiple attempts at repression by Canadian police and military. According to Tawinikay, a Michif-Cree two spirit person active in the struggle against colonialism, resistance only spread: "it began when the matriarchs at Unist'ot'en burned the Canadian flag and declared reconciliation dead. Like wildfire, it swept through the hearts of youth across the territories. Out of their mouths, with teeth bared, they echoed back: reconciliation is dead! reconciliation is dead!"[17]

Parallel to the camp at Unist'ot'en, the Tiny House Warriors erected a series of small houses at strategic points in Secwepemc territory to block the Trans Mountain pipeline "to assert Secwepemc Law and jurisdiction and block access to this pipeline." Their actions simultaneously halt the expansion of the dying world of colonialism and fossil fuel capitalism, and plant the seeds for a healthier world by "re-establishing village sites and asserting our authority over our unceded Territories. Each tiny house will provide housing to Secwepemc families facing a housing crisis due to deliberate colonial impoverishment" and be equipped with "off-the-grid solar power." Racist settlers have attacked and robbed a Tiny House Warriors camp, and several members of the group face charges for land defense actions.[18]

After police raids in February 2020, Indigenous people and supporters "Shut Down Canada" with rail and highway blockades across the continent, bringing the Canadian economy to a halt. Through March, there were blockades and heavy protests in territories occupied by the provinces of British Columbia, Ontario, and Quebec. Gord Hill has emphasized the Indigenous practice of setting up solidarity blockades across the continent, a major occurrence in 2006 as well as during resistance at Oka in 1990.[19]

It is a particularly effective model for resistance, showing how anticolonial warriors can challenge both the legitimacy and the logistical viability of a settler state that is fundamentally based on extractivism, and how struggles that are intensely local—and frequently dismissed by the mainstream as small, backwards, or irrelevant—can in fact very quickly scale up and take on continental dimensions.

Similarly, we have the case of the airport blockades that shut down hundreds of airports large and small across the US and immediately forced the government to rescind the Trump administration's racist "Muslim Ban." This tactic shows how effective decentralized networks can be, bringing the country to a halt and threatening an infrastructure that is crucial to the fossil fuel economy and to the inherently racist border regime that punishes people who are fleeing the effects of US economic and military actions.[20]

While we are still waiting for the tactic of airport blockades to make a reappearance, the resistance of the Wet'suwet'en, as well as that of the Lakota and Dakota at Standing Rock, has led to a proliferation of similar struggles targeting the Bayou Bridge pipeline in Louisiana, the Trans Mountain pipeline going from Alberta to British Columbia, Enbridge Line 3 running from Alberta to Wisconsin, and the Mountain Valley pipeline from West Virginia to Virginia. Meanwhile, other pipeline projects were canceled due to investor fears of how resistance could affect profitability.[21] There are fewer pipelines in the world today thanks to these movements.

Aside from blocking destructive projects and policies, direct action can also achieve the protections that government regulations promise and fail to provide. The UK's marine sanctuaries are an infamously bad joke, providing almost none of the promised protections to marine life. In response, Greenpeace took out a boat and dumped several large boulders into the sanctuary, ensuring that any dragnet fishing trawlers that came through would wreck their own equipment.[22] They effectively used the threat of property destruction to prevent the ecocidal stripping of the sea floor. This practice is significantly more widespread than most people know. Visiting friends on an island on the Dalmatian coast of Croatia, where much of the older generation participated in the partisan resistance against the Nazis, I chanced upon the interesting story of how a neighbor dumped an old junker—engine removed—into the middle of the small bay where they lived, sabotaging any larger fishing trawler that came in to drag the waters. It is unknown how many bays are similarly protected throughout the Mediterranean and its adjoining seas; neighbors acting autonomously and sometimes illegally rarely announce their actions to the media or to experts. The bottom line is that nature sanctuaries declared by governments are illusory. The only thing that makes a place safe from the institutions—public and private—that would despoil it is a certain correlation of force that favors local people.

FOOD SOVEREIGNTY AND ECOLOGICAL HEALING: FINDING OUR PLACE IN A DAMAGED ECOLOGY

Industrial agriculture has failed. The so-called Green Revolution pushed by institutions like the World Bank and agrochemical corporations has not succeeded in ending hunger, which is largely a product of colonial and neocolonial systems for food production and distribution.[23] It has succeeded at its real objective, transferring a huge quantity of wealth to private companies and investors, while making rural areas more legible to state intervention. It has constituted a major land grab, with one percent of farms controlling 70 percent of farmland worldwide,[24] its disrespectful practices are destroying the soil, with one third of the soil of the entire planet "acutely degraded,"[25] and it is driving global warming. Dependent on heavy machinery, poisonous chemicals, and gratuitous global supply chains, industrial agriculture is directly responsible for 14 percent of greenhouse gas emissions, and is also a major driver of deforestation, responsible for another 18 percent.[26]

But attempts to heal and recover the land, to gain local control over food, are serving as a springboard for some of the most exciting resistance and the most intelligent alternatives in this terrain.

One such experience has had a direct impact on this book. The neighborhood where I currently live is surrounded on three sides by agricultural land that has been left fallow for years. In fact, the fields are a great example of the uselessness of state environmental regulations. To carry out the more profitable industrial animal farming (in which animals are imprisoned in wretched warehouses for their entire lives and fed on a diet of grains that makes up a large part of overall agricultural production, the complete opposite of traditional pastoralism), the state requires farmers to have a certain amount of land, supposedly to offset the impact of the animals. Without planting anything, the tractors come through every year or two and plow the fields, wasting fuel, killing the soil, and preventing the fields from becoming a carbon sink and a habitat for grassland species. All to fulfill a bureaucratic environmental regulation.

Several neighbors, myself included, squat part of these fields in order to garden. In my garden, I combine fruit and nut trees with some food crops and medicinal herbs, use rainwater collection, and don't use any chemicals or motorized tilling. It's a small garden and though I don't rely on it for the majority of my diet, it does provide me with healthy food, a way to learn from the earth, to listen to what's happening, and to learn methods for gardening in a desertifying climate, which will be a major issue here over the next decades and one that mainstream producers are completely unprepared to deal with. I have been able to see, in just six years, how the soil health has improved dramatically, increasing water and carbon retention and providing a home for so many other species. And I share the medicinal herbs I grow in a network that brings together anarchist infrastructure projects throughout Catalunya, fomenting everything from farms to print shops, pirate radios, and free schools while fostering a gift economy among them.[27]

Beyond the neighborhood, there are extensive fields of olive trees, most of which have been abandoned by their owners. Many of these orchards are now taken care of, from pruning to harvest, by neighbors who organize themselves and win access to healthy, high-quality olive oil that the capitalist system would normally reserve for wealthy consumers.

All of this is only possible because of our radical disrespect for property laws. In some cases, people ask for permission: a few of the absentee owners are simply small holders who are too old to do the harvests themselves, but in the last analysis, no one would respect the property rights of an owner who insisted on letting the bounty go to waste.

Teia dos Povos, Bahia, Brazil

To help out with this book, anarchist comrades in Brazil spoke with Erahsto Felício of the Articulação TP, the communication section of the Teia dos Povos network.

> Teia dos Povos is an articulation of many base nuclei (territories) spread throughout Bahia and other states. Most of the territories were taken back, which means they do not have a long history of organized occupation. In the state of Bahia, most of them have been occupied in the last 30 years. Those with a historical occupation, such as some *quilombos* and Indigenous territories, have gone through an organization process and sometimes self-demarcation in the last few decades. [*Quilombos* are maroon communities of escaped Africans and other refugees from colonization. Self-demarcation is an Indigenous community's declaration of its own territories.]
>
> Currently, most of these territories live from what they grow. They are, therefore, producers and generally agroecological or in transition to agroecology. The territories are located in different biomes such as the Atlantic Forest (where Teia dos Povos originated), the Caatinga (a dry forest ecoregion in the northeast) and the *restingas* (sandspit or coastal broadleaf forests on sandy soil) and mangroves. There are some traditional territories in which life is still associated with collecting, fishing, and small farming, yet there are others that even have a fine chocolate industry. We pursue food sovereignty whenever possible, as well as other autonomies.
>
> Our main conflicts are with mega-enterprises. On the coast, these are linked to shrimp farming and tourism, especially hotels. Inland, it ranges from agribusiness estates to mining companies. In the urban territories, peripheries, and homeless settlements the major conflict is with the state and its police, whose records show that 97 percent of those killed by the police in Bahia just in 2020 were Black people, that is, there is a black genocide under way in our lands.
>
> The nature of the attacks on the territories varies from region to region. In the west we have seen agribusiness dry up entire rivers for their production for the foreign market. Now we are seeing this same operation taking place in Chapada Diamantina, where landowners obtain water grants to the point of making the water supply of many peoples unfeasible. There is also the harassment and persecution per-

petrated by local politicians. Violence is frequently linked to the power of large landed estates or to the destructive exploitation of the land (mining and the use of river water for large-scale irrigation). Mining companies usually have foreign capital while agribusiness estates still belong to national capitalists, but they have been counting more and more on external financial support.

Resistance has been established according to the principles of food sovereignty and the massification of the struggle. Our elders have been teaching us that the land is the foundation, the principle, it's where the great struggle is born. And we saw that many people could not resist harassment by corporations, as they turned up distributing food parcels, jobs, and some other "charities" in order to dismantle the movements. How do we face this situation? For us, it's through the land, by creating food sovereignty, real abundance, a small paradise of abundance. As a result, coercion would have much less impact. The other matter is the making of alliances. Nobody can win alone because of the immense power of capital. Thus, a quilombo under attack needs support from Indigenous people or from MST settlements [the Landless Workers' Movement] in the region. When the COVID-19 pandemic began, for example, there was a lot of hunger in the suburbs and within the prison system, and our comrades from the Política Reaja Ou Será Mort@ ["React or Get Killed"] Organization campaigned to get food and hygiene products for prisoners who were banned from receiving visits from family members and, therefore, lost access to much of the food they received. Then an MST Brigade connected to Teia dos Povos went to Salvador (the capital of Bahia) and sent a ton and a half of organic food so that they could reach the prisoners' relatives and the prisoners themselves.

So we learned that it is essential to establish the Black, Indigenous, and Popular Alliance in practice. Without it, it is impossible to resist. And, mainly, that the resistance comes from the land that we are reclaiming.

We usually say that we are heirs to a tradition of long-standing struggle in the history of Brazil. If we look closely, the struggle in Quilombo dos Palmares lasted longer than the Soviet Union's socialist experience. Palmares faced the main economic powers of its time: Portugal and Holland. It managed to establish itself as a federation of peoples under the leadership of the Bantu peoples, but which had Indigenous peoples and even Jews and Muslims persecuted by the Inquisition among them. This is just one example. We could talk about Cabanagem, Balaiada, and the city of Canudos, located in the hinterland of Bahia, which faced the Brazilian army with important victories against the newborn Republic of Brazil.

We identify a rebellious tradition of peoples and communities (not individuals) who joined forces in a fight against the latifundium—our longest-running enemy in these lands. So the Brazilian rebellions were built by an alliance of entire communities and peoples who rose up against large estates and even the power of the state. We see this throughout colonization, through the nineteenth century until the beginning of the twentieth century. It was in the twentieth century that political theories started giving more focus to the organization of individuals through political organizations and parties, abandoning the perspective that it was fundamental to territorialize the struggle through the adherence of communities to the rebel march. We are not doing anything new, just learning from our elders and those older than our elders (the enchanted ones and history).

Taking back the land is the main task at the beginning of the rebel journey. The whole political theory of the Brazilian peasantry is all about breaking the fence and getting inside. We are wondering what to do beyond the fence. How to build the territory. When we return to a land degraded by cattle raising or by the exhaustion of the soil by crops grown without considering the health of the land, it is necessary to begin a process of regeneration. But we know that the people will only stay on that land if we have food. So the work starts first. Before we even break through the fence to retake the land, we plant in other areas to have enough food for three to six months for the people who will retake the land.

Accumulating food is essential to the process of taking back the land. Without food, there is a risk of taking back a degraded land and people going hungry and giving up on the dream of the land quickly. But if we have food for these first moments and break the fence, then we need to make the first garden to produce an amount sufficient for our existence: beans, sweet potatoes, and corn (three months), cassava (six months), and banana (one year) are species that we always plant because of their ability to generate a healthy and rich diet.

While we are making this garden, we need to work hard to raise our agroforestry. For this, we have a seedling nursery with native and fruit trees (açaí, cupuaçu, cocoa, etc.). Growing the forest is the beginning of the land regeneration process. We developed a four-hectare project that is our best reported experience to be able to share with our companions in land reclamation. All of this respecting the principles of agroecology that we learned from the native peoples and intellectuals like Ana Primavesi.

Teia appeared in 2012—the year of the end of the world in the Mayan calendar, the year of the Zapatista silent march—in the Terra Vista Settlement (municipality of Arataca, southern Bahia) on the first Agroecology Day in Bahia, in articulation with native nations like Pataxó, Pataxó Hã-hã-hãe, Tupinambá, quilombos, and peasant movements like the MST. There, a struggle was launched for the spread of agroecology and the spread of creole seeds [local traditional varieties free of intellectual property regimes] in the territories. It started as it should: with the seeds. It was from the spread of creole seeds of corn and beans that the network was woven. As an articulation [in contrast to a centralized, formal organization], we have no role in directing the territories. On the contrary, we want to learn from the differences in direction, work, and struggle in each territory. We are about to build bridges between different territories and to build the necessary alliance to face racism, capitalism, and patriarchy. Although there is no board of directors, there is a council of older people willing to guide struggles, but the decision is made by each base group. We are spread among these basic nuclei (territories, peoples, and organizations that are territorialized) and the links in the web (supporters, diffuse urban groups, political or research collectives). So the links need to support the struggle of the nuclei, but only these can lead the fight, only the nuclei can guide the direction of Teia dos Povos. Because we assume that only those who hold a community, only those who broke the fence, occupied the land, can explain how they do it. Those who have not yet territorialized and organized their community cannot lead a fight for land and territory!

On our last journey, in 2019, in the heart of the Payayá people's lands, in Utinga in Chapada Diamantina in Bahia, a comrade from *Rádio Zapatista* was with us to discuss the life and struggle of the indigenous Zapatistas from Chiapas. There is no doubt that there is a deep inspiration in the struggles of native peoples in search of autonomy such as this Mexican comrade or even the Colombian Guards or the Mapuche people in Chile. They are stories and struggles that feed back into what we do. Almost all of these struggles we mentioned are fundamental because they combined the reclaiming of the land with the awareness that it was necessary to help nature to recover. They have a deep love for humanity, but also for *Madre Tierra*, as they say. Now we also hear that there is an echo of these struggles in Rojava, Syrian Kurdistan, and we are very happy. We hope one day to get to know them deeply. For us, the path is the defense of territories and peoples, above all recovering every inch of land that belongs to the native peoples. There is a debt to be remedied on this issue. It is no longer possible to tolerate entire Indigenous nations being deterritorialized. There is a climate threat now and we all know that Indigenous peoples are the greatest guardians of the forests, there is an urgent need to return the land to the native brothers and sisters.

The Left needs to rethink with some urgency the materialist and individualist horizon that they got themselves into. It is essential to understand that a true alliance with Black and Indigenous peoples also involves recognizing their worldviews. Defending a river is not just preserving a natural resource or a good deed for the environment. For the Nagô people who came from the region of Nigeria, the river is the representation of Oxum, a deity. When we desecrate the river as a divinity and make it a natural resource, we convert it into merchandise that can be bought, sold, destroyed.

In the same way, Indigenous peoples understand that their enchanted people, spirits, and guides who care for and sustain the world live in the forest. They do not only see minerals, wood, and animals there. These worldviews are fundamental to safeguard the world in collapse today. So without abandoning stupid materialism, it is very difficult to move forward in a true alliance. On the other hand, it is necessary to reject this individualism that no longer speaks to peoples and communities. It is necessary to go beyond the congregation of individuals under ideological banners. We want to know if X people or neighborhood Y can get together to discuss reclaiming land, planting trees, building organized territories where we take care of each other. And that means building another time for politics. A time that is aware that it will take fifty years for a baobá to give its seeds for the next generation to be able to plant. Our problems and sorrows are urgent, but the haste with which we are acting in politics has simply not helped us to build up politically and overcome our enemies. It is necessary to cultivate a good individuality so that we have more solid collectives. And that good individuality is the self-care, the self-preservation of your people, it is to cultivate values that connect you to your people, to your ancestry, to your spirituality in the territory you live in or that was taken from your people generations ago.

Finally, the oppressed peoples of the world recognize each other in the first gesture of struggle and are willing to make alliances. We just have to examine our differences and work on what brings us together. There are two words of wisdom that we echo at the end of our writings and that we wanted to share here: What unites us is greater than what separates us; Peace among us, war on tyrants.

Guaraní, Brazil

Another inspiring struggle taking place in Brazil has brought together Guaraní communities recovering their lands and a rural anarchist collective called Cultive Resistência. Comrades helping out with this book passed my questions on to Catarina Nimbopuruá and Aldeia Tapirema in the Piaçaguera Indigenous Land, Peruíbe, Sãp Paulo.

The lands where I always lived were in the Atlantic Forest (Mata Atlântica). I was born close to the hills, there was a very beautiful waterfall, a very beautiful forest and if we planted there, it would give a lot. But today I am on the coast, on the beach, where the land was also very good, until the mining company and other forces came here and finished off the forest. But we are still trying to plant again. For me the forest is very precious, even more so being the Atlantic Forest.

Our lands were devastated by the mining company. But today, we are very happy here because we returned to the land that once belonged to our ancestors. So, for us it is very important because here is the spiritual strength, the spirituality of the elders, the ancestors. They are here. And that is why for me this land is very valuable. And also because it is the Atlantic Forest, there are many birds that stay here, there are several different fish, there are turtles around here, in the sea. So, here for me it is very special, this land here for me is very, very special.

The resistance of the Tupi-Guarani people goes back to 1500, five centuries ago. Our people have been resisting to preserve and maintain the culture, our customs, songs, dance, and especially our mother tongue. The advancement of technology is leaving many young people distant from their culture, and today we are focused on strengthening the culture and the mother language.

We do classes in Tupi-Guarani (the language) and gather around the campfire with *txeramoî*, *txedjaryi* to strengthen our knowledge, through lived and everyday stories. We build *pau a pique* houses, we harvest herbs to make medicinal tea. [Translator's note: *txeramoî* and *txedjaryi*, masculine and feminine, are the wise elders who concern themselves with spirituality and healing among the Guaranis. *Pau a pique* is a bamboo and cob construction technique.]

There are several examples of struggle and inspiration, such as the struggle of our elders to retake old lands and the struggle for their demarcation as Indigenous territory. To have autonomy within the community itself, such as differentiated education and traditional healthcare.

Our land was very devastated by the mining company, the land was very, very harmed. But even so today we are taking care of it, so that we can plant what we need for our daily food like sweet potatoes, cassava, and fruit trees. We are taking care of the land and I imagine our mother earth is very happy at the moment, because we are taking care of it. We are taking care of our children's future.

It is very difficult to recover our land today due to government laws. When we go to recover a piece of land we unite txeramoî, txedjaryi, children, and women, and we set up camp in that place and fight against the government that insists on not demarcating our lands. To heal the degraded areas we do a study on the place, like what kind of animals live in that environment and we plant trees that the animals feed on.

We have gotten support from Cultive Resistência. Our relationship with Cultive Resistência started a long time ago. We started a friendship and work relationship in the Piaçaguera village. This is a partnership that has been generating great efforts to strengthen the Tupi-Guarani culture, and this relationship was and is fundamental for our community to resume practices and customs that were only in the memory of the elders. With the work developed, our children began to feel proud and appreciate their own culture. Since then, this relationship has been a source of great pride for our community.

Plan Pueblo a Pueblo, Venezuela

"If food is a human right, our goal is abundance and diversity. If food is a commodity, it is claimed by the logic of scarcity."

Ana Felicien works in a research institute in Venezuela and collaborates with a Latin American working group developing the concept of political agroecology. "One theme we want to center is how the nucleus of agroecological thought in Latin America is not the academy, but all the histories and the construction of Indigenous, Afro, and peasant peoples." She has participated in various agroecological initiatives in urban and rural spaces, including la Universidad Indígena del Tauca (the Indigenous University of Tauca) and el Plan Pueblo a Pueblo (Plan People to People).

One of the Plan's triumphant moments was when they were able to secure funding from the government program to provide meals in schools in rural zones and some cities, and were able to dramatically improve the quality and abundance of the meals for children. "With the budget that had been spent on feeding 400 children a diet of just four crops," Ana explains, "the Plan managed to feed 8000 to 9000 children a diet of 27 different crops. So basically we were showing how obscene the conventional system of production was. And so they canceled the program."

These initiatives are developing in a situation of extreme economic and political violence, between the financial blockade orchestrated by the US and its allies, and paramilitary attacks that correspond with elite economic interests.

> Plan Pueblo a Pueblo is the articulation of a plan between the countryside and the city, it is not an organization. The plan was put forth by a group of *compañeras y compañeros* in the midst of the crisis we are experiencing, in which food has become a central topic of our lives. That's to say, because of inflation and the high prices of food, people have to plan and struggle every day to get what they are going to eat, and in this situation we were able to build this proposal to connect rural communities and the city in order to carry out the planning of production and consumption and from there to think of how to go about decolonizing our forms of consumption. In the city, our consumption is very much governed by a few principal crops, very globalized, with imported seeds, mostly from the United States, with lots of agrochemicals, and that agricultural model have to be cultivated by peasant families in the places with the better resources for production (most water, better lands, etc.). And in the midst of the crisis, we managed to deploy a proposal for the agroecological transi-

tion towards food sovereignty, which gets talked about a lot, but which is very difficult to build in a particular terrain with people.

Meanwhile, some of us were also a part of the process that we called the Popular Constituent Debate for the Seeds Law, from 2012 to 2015, something we did on a national scale to try to create a law that would recognize all of this—peasant knowledge, the importance of our diversity, custody of the land and the real ways of practicing an agro-ecological agriculture. Because despite all the resources that have been spent throughout agricultural modernization, in the end most agriculture is cared for by people of peasant origins, Indigenous origins, African origin, and these are the most widely grown and most available crops.

The hunger crisis in Venezuela, which is publicized so widely in the media, is that there aren't any more imported foods, but every time we're in the countryside, people are saying: "there are too many root crops, but we don't know how to eat them." There, the problem is abundance. So it's really schizophrenic that you have the image of a Venezuela in famine and then a Venezuela where food is being wasted because "it's not food," because it's the food of Black people, food for Indians.

Pueblo a Pueblo arose in 2015, which was a very hard year, because our diets are very monopolized and homogenized. Basically, we get fed on very few crops, above all processed food. The most consumed food is processed corn flour, and this has repercussions throughout the entire agricultural system. It's controlled by a single company, Empresas Polar, only one type of flour is produced, from processed white corn, and nearly 80 percent of arable lands are planted with corn and rice. And this corn, basically 90 percent of it is seeds that are hybrid varieties oriented towards industrial agriculture. So we have an enormous diversity that is completely invisible, and all the resources for production and the structure of the agroindustrial system is centered on these few hybrids, which aside from being imported give lower yields under our field conditions, to name just a few of their characteristics. So the average yield of the corn is 3,500 to 5,000 kilos per hectare, which is very low. In the United States, it's two or three times as high. But the yield of cassava or other root crops, yams, sweet potatoes, etc. can reach 20,000 kilos per hectare. But the majority of the land is planted using imported corn seeds, which monopolizes agricultural policy, resources, and so on. Ninety percent of the people live in the city, as the result of a migratory process that we can trace through history, of going to the city because I can achieve a higher standard of living in the middle of what was called the oil boom.

Some *compañeros* who came from the historic struggles for the land in the '70s and '80s, in the student movement, who were no longer in the state institutions because of differences of vision, took on the construction of Plan Pueblo a Pueblo on the premise that food was a right and not a commodity. So they went to Carache. This is a city in the Venezuelan Andes, which is the region with the highest crop production in the country, and they went to rebuild what they call the struggle of Argimiro Gabaldón, who was one of the leaders of the *guerrilla* in Venezuela and who carried out an interesting labor for peasant organization, for self-organization.

They began to recognize different axes of production, different routes, because in the Andes, in the middle of the mountains, production is mostly done at the family scale. So what you have is a very intense production which is oriented towards supplying national markets, but on the basis of family production, with a long history of cooperative organization. In other words, all those highways were built by the people. There's a long history of work exchanges, *cayapas*, *mano vueltas* [respectively, a form of collective work in which a group carries out a task one person alone could not, and a form of reciprocity in which one day, you and I work in your garden or on your construction or other project, and the next day we work in mine], family relationships which are like *medianerías* [a juridical and architectural situation in which multiple houses or families share an overall structure that is subdivided by internal walls between the different units] but which are really relations of sharing and complementarity among families. Which are also bisected by other power relations. For example, the majority of women are day laborers, and the men are the ones who own the land, because they are the ones who can get investments.

So, considering this to be a strategic location, because, also, in terms of geography, it's quite strategic, an analysis was carried out with these cultivators to create Plan Pueblo a Pueblo. The goal was to generate a response from below to this assault, this idea that food was a commodity and if we didn't behave, we couldn't have any. Ours was a response that said: "food is not a commodity, it is a right, and it must be in our hands, in the fields and in the city." The essence of the plan is what we call the ladder method for double participation. People organize themselves in the fields to produce and to establish a calculation for prices on the basis of costs and the needs of people in the city, and in the city people identify their dietary needs.

There has to be a connection between every rung on the ladder, between the countryside and the city, so that people in the countryside can plan on the basis of a diagnostic of the needs of the city. These needs are enunciated by their consumption patterns. Another objective is to fit what is produced with what is consumed. We have a diversity there that does not enter into the distribution circuits, but they produce so much and people don't eat it because they don't recognize it or because they are very domesticated by this culture of supermarkets.

The Plan has managed to distribute more than two million kilos of crops, I've already lost count. We have established different points for communal distribution, where people plan out their orders on the basis of the community's needs. The neighbors of Carache pick up the produce that has been ordered and they do distribution days in the urban zones, and they pay the farmers the same week.

We organize the cultivation of 27 different crops, many of them traditional, some more conventional.

The greens, the leafy crops, are grown primarily by women. An interesting thing about the Plan was that these territories have a long history of cooperativism, but who participates in the cooperatives? Above all men with a medium or high production. But then you had all these people who were much farther from these official productive axes, with poorer roads and smaller fields. The Plan gathered together the very smallest. "You just have a little cilantro in your front garden? We'll take it. Anything you have, we'll take it."

This generated a very interesting dynamic. You had women who began to fight for the land, to ask their husbands for more land, their daughters began to get involved. If you had a family that wouldn't pay the student housing for the daughters in the city, once the mother and the daughters were working the land and had their own money, all of a sudden it was a priority for the daughters to go to the university and they could pay. All this was possible because the productive geography was changed. Our distribution truck passed through the most isolated places.

All this went into crisis once there was no more fuel for driving our routes. 2020, above all, was a very hard year. The shortage of fuel

> affected how we calculated our prices and everything. But if food is a right, then everything you need to produce the food is also a right.
>
> This also means seeds. If a good seed is one that maximizes production, even if it costs $400 a pound, that's not sustainable when the minimum wage is $1 a month. That's 400 minimum wages. So our task now is to intervene in all the elements of the forms of production and move towards an agroecological model.
>
> In many territories, the rains are changing, we're already facing climate chaos. The agroindustrial system is in crisis, they don't have gasoline to fumigate the corn fields, which is usually done by airplane. So there has been this great wager on decentralization. People are beginning to sow to fields, reconnecting with a tradition that synthesizes what has survived from the African, the Indigenous, and the peasant traditions.
>
> The difference between the earlier crisis and 2020 is that now people realize there is the possibility to supply themselves with their own efforts. Everyone is cultivating. And the other half are processing and exploring a complementary transformation of their consumption. And the old seeds that people had been saving have reappeared. Now there are nuclei for the production of seeds all across the country, fairs in the villages for sharing seeds, varieties, experiences. Opposed to the logic of dependence. The only possibility to eat with dignity is by forming a part of this network that already exists.

Food sovereignty is an important practice for ending white supremacy, in the Global North as well as the Global South. Leah Penniman, who participates in the Soul Fire Farm outside Albany, New York, and is the author of *Farming While Black*, identifies the dispossession of Black farmers and "food apartheid," a racially distributed lack of access to healthy foods that results in a huge number of deaths due to diabetes, kidney failure, and heart disease, as ongoing features of systemic racism. Whereas "farming is inherently about the future" because of the planning, the long-term perspective involved in planting a tree that won't bear fruit for several years, many young Black people have a sense that there is no future because "incarceration and untimely death are so ubiquitous."

Addressing these overlapping concerns, Soul Fire Farm trains farmers of color, practices silvopasture—grazing birds and sheep among fruit trees as a way to maximize carbon sequestration—and donates lots of fresh food to neighborhoods in Albany that are victims of food apartheid. In Penniman's words: "We use Afro-indigenous and regenerative practices—fancy words that essentially mean we're trying to farm using the best advice of our ancestors and we're trying to farm in a way that actually makes the environment better and not worse."[28]

Inside and outside of the cities, Indigenous peoples have been at the forefront of the struggle for food sovereignty or food autonomy. In rural areas of North America, this has often centered on struggles for traditional hunting and fishing practices. Opposition to native hunting and fishing often comes from commercial fishers under the pressure of producing for the capitalist market—a practice that has destroyed marine ecosystems around the world—and from reactionary whites threatened by the implication that they live on stolen land.

Native hunting and fishing practices are very much about caring for the ecosystem as an integral part of the territory, rather than as an outside agent. Angela James, a hunter from the Bigstone Cree Nation, explains how: "the moose, the bear, the elk, the muskrat, the fish, all these animals, these beings, they're our relatives. You've got to honour that protocol, honour that connection, that we are part of something bigger than all of us, we aren't almighty human beings at the top of the food chain, 'cause we're not."[29]

Pastoralism from Barcelona to the Pyrenees, Catalunya

Edu Balsells practices traditional pastoralism in Catalunya. A participant in the autonomous movements, he began to get involved with the world of activist agriculture in 2004 and has since become a principal reference for the use of goats and sheep in forest maintenance and the prevention of catastrophic wildfires. He also participates in the transhumance, a semi-nomadic practice going back to the beginnings of pastoralism, in which the shepherds stay with their flock in the lowlands during the winter, and then migrate to mountain pastures during the summer, allowing for a healthy regeneration of vegetation in both places. The Pyrenees and their various lowlands are one of the few places in the Global North where the transhumance is still practiced. The shepherds who carry it out have long fought for the commoning of land, whereas its opponents are generally those who profit off the enclosure of lands and the industrialization of food production, or those who, in terms of public order, oppose public spaces being used jointly by humans and other animals.

Edu currently has a small flock of 150 sheep, a sheep dog to guide the flock and a mastiff to protect it, all of traditional Pyrenees breeds. We met on the small terrain where Edu has his cabin and the open stable where the flock spend the night when they're not out on the field. It is on the slopes of Sant Ramón, a small mountain on the outskirts of Sant Boi de Llobregat, itself a small city within the Barcelona metropolitan area. He has been there for ten years.

Edu's income derives from selling lamb and selling manure. The latter is actually illegal without a permit that is too expensive for small producers, though he defiantly told me to mention his transgression in the interview. Another source of income is a government fund to maintain the forests and reduce fire hazards, technically a subsidy, though the government would have to pay for the service either way, and forest maintenance with machines costs 50 percent more. Unfortunately, Edu cannot sell the wool from his sheep because the price for wool that is off white or spotted does not even cover the cost of bringing it to market.

Edu began by explaining to me the importance of the traditional animal breeds in the kind of pastoralism he practices:

> The first two years we were here with goats. Of course every animal has adapted to a certain behavior for its survival. Cows eat as though their tongue were a scythe, goats raise their heads and look for the tender shoots on the trees, sheep tend to eat the herbs and grasses, horses have teeth on top and on bottom and they can rip out the vegetation that a sheep or a cow couldn't, so every animal has adapted. The goat is, *par excellence*, the animal best suited to forest pastures, without any doubt, it is a self-sufficient animal that can get one hundred percent of its diet from the forest. The problem is, when we came here to Sant Boi, even though the land here is forested, Mediterranean forest, we realized that "la cabra tira pel monte" [an adage that every being will follow its own nature, but that literally means "the goat heads for the hills"]. The goat is a very selective animal, it walks a lot, and here the space is limited. This isn't a mountain of 600 hectares, the forest we take care of is only 60 hectares and that comes with a lot of people, trails, it is a space that is very fragmented by suburbs, electrical towers, people doing outdoor activities, so here the goats were not working out as well as they could have. And the space that we're supposed to manage is a margin of 25 meters around the outlying residential areas. The city pays us to protect the suburbs from the forest but as far as I'm

concerned, we're protecting the forest from the suburbs. But we have to keep the space clean, with a holly oak here, a bush there, very few grasses.

But goats are very aggressive in these spaces, they would go after the trees and they wouldn't touch the grasses. So they were always in the forest, which is fine, but we're paid to carry out a very technical labor to keep the brush down and prevent fires.

The effect of the goats on the forest was huge. Already after one year, people were telling us, where the goats had passed, now there was more space, trails were opening up, there was wild asparagus, mushrooms. Of course, it also depends on not saturating a space with too many animals.

If we don't take care of the forests we might lose it all here in Catalunya. There have even been forest fires in the Pyrenees in the winter. That's unheard of. Between heat waves and fires, we might not be able to practice pastoralism or agriculture anymore. But if we support grazing in the forests, that would open them up and allow more light to come in, better water retention. We could also use the energy resources of the forests like biomass, with local communities collecting fuels, not industrially which would destroy the forests, but done with practices that emphasize balance and sustainability. You have to make sure the trees and bushes can grow, the birds can nest, the boars can forage. There are small activities that would make a huge difference, but in the current system it is not economically feasible.

So we were with goats for two years and we had to change. If the type of pasture that you have is more suited to sheep than to goats, then you change. With all the heartbreak in the world because they're your animals and they're used to you, but in the end you have to make a living. They live off of you and you live off of them. It's a balance.

The forests are as bad as they are because of a lack of interaction and maintenance by humans. Because we think we're a very rich country, and so we don't value the energy and the potential that the forests have. On the other hand, if Repsol [a major energy company] came and said, these pines are worth 100 euros, there wouldn't be any pines left in all Catalunya. It always goes from one extreme to the other. In the '60s they got rid of the goats, and then around 2000 they said, well, maybe we can integrate goats in forest maintenance. But they never speak with the shepherds, to find out what our proposals are. We tell them, but nobody listens. And then somebody comes from the government or the NGOs, all of them living off of public money, and they bring their proposals. It's like, *you just don't get it!*

We don't have any remedy to the global market we live in, so we have had to live with getting subsidized and to accept that they pay us and we're obliged to do certain things to get that money. I did calculations earlier and if I just made two euros more off of lamb [per kilo], I wouldn't have to receive subsidies. But it's the global market that regulates prices. Sure there's a market here, Sant Boi has 80,000 inhabitants, but the market is organized in such a way that meat comes all the way here from Mexico or Chile, and then all the lamb from here gets loaded up on boats in Tarragona and shipped to Syria. Because they eat a lot of lamb in Muslim countries and there's a war over there, so they can't produce it themselves, they ship it from here. So, what? We're supposed to be thankful that there's a war in Syria so that lamb producers here can make a living? Well what a shitty system, then. And that's how it is.

If, together, in an organized way, there were more unity in our sector, if we were capable of making ourselves known, proposing and offering and valuing our function in this world, in terms of the environment, culture, maintaining traditional breeds and practices, encouraging biodiversity, so that we could actually make a living from what we do, then we could give a good kick in the ass to the Department of Agriculture, the European Commission, the Spanish Ministry, and then send them on their way. But we have been unable.

It's very easy for governments to talk. They only know how to lie. It's one lie after another and I've seen quite a few official projects to integrate shepherds into forest management. But if you can't guarantee that the shepherd can make a living, nothing will work. Everyone has to eat.

There should be an exchange. I don't mean a monetary one. I have sheep, we could do an exchange for cooking oil, gasoline, vegetables, whatever you want, something to guarantee our survival.

It's very easy for governments and all these satellite organizations, from environmentalists to unions to the hunting lobby, a whole conglomerate of organizations who make a living off of talking about the benefits of pastoralism for the environment, for rural development, and it's all true, but for all they talk about it, they don't accomplish anything. All of these bureaucrats and public servants, they're always talking about the numbers, well let's look at the numbers. Agrarian incomes are in free fall, the prices of what we produce are in free fall, but the cost of living is rising. The cost of rent in rural areas is going up like it's Barcelona. In a village with 12 inhabitants, that doesn't even have a bakery, a pharmacy, or a bar, rent is 400 euros for an apartment. I'm not paying for the bourgie lifestyle of living in the countryside so I can telecommute.

There's a huge pot of money that is earmarked for projects like mine, but coincidentally the shepherds barely see any of it. First they have to pay for a study, then a pilot project, then a diagnostic, then an analysis, and there are all these people who really care about environmentalism and food sovereignty and they're on your side, but in the end, everything they've done hasn't helped me out at all, things are just getting worse. There are more and more people talking about these things, but they're not getting any better.

Now I just don't speak with academics when they call me for interviews. Because they exploit us, they sell knowledge and don't give us anything in return.

But I don't want the shepherds to disappear. Our world has a lot of history. The pastoral world has these connections that surpass borders.

As I finish up this manuscript, the tenth anniversary of a great victory at Cherán K'eri has come around. On April 15, 2011, the people of this small town in the Mexican state of Michoacán rose up to defend their forests, their water, and their lives. Cherán K'eri, with a population of 14,000, is one of the principal towns in the territory of the P'urépecha people. Thanks to the last hundred years and more of struggles by Indigenous peoples from Baja California to Chiapas, large swathes of territory throughout Mexico are officially recognized as communal lands, including 15,000 hectares around Cherán K'eri. However, nothing is safe under capitalism, and much of the communal lands were being despoiled by the drug cartels, which are largely integrated into the state and which have diversified into other industries like lumber.

Several men in the town had spoken up against the out-of-control logging, and they usually ended up dead. As the killings continued, unpunished, and as the logging approached the source of the town's water, the women rose up and took several cartel truck drivers and loggers hostage. There were several days of intense fighting against the cartel's mercenaries and the local police, but the people of Cherán K'eri put up barricades, set fire to trucks, and held their own with stones,

Molotov cocktails, and whatever firearms they could get their hands on. On April 17, they created a "popular assembly" that would be the first step towards their self-government. From the assembly arose a dialogue commission consisting of rotating representatives from each neighborhood. This structure served the egalitarian aspirations of the people of the town, and it was also far more effective than having leaders who could be co-opted, kidnapped, or assassinated.[30]

Around the barricades and the *parhankua*, the communal cooking fires, a sense of community was rekindled, overcoming divisions, antagonisms, and scarcities implanted through hundreds of years of colonialism. P'urépecha traditions and language were revitalized and became a cornerstone of their practice of autonomy. One such tradition was *kuájpekurhikua*, a word that translates as "taking care of the territory" and that refers indistinguishably to the social and ecological territory, therefore including everything from education and improving the situation of women in the community, to repairing relations between neighbors, to massive efforts at reforestation. By 2015, the nursery they established for growing trees—starting out with seeds they had gathered in the forest just four years earlier—surpassed the figure of one million trees and shrubs germinated a year, with an 80 percent survival rate, making it the largest greenhouse in the state and possibly in all of Mexico. The people of Cherán K'eri also developed a communal justice system focusing on mediation rather than punishment. By winning their autonomy from the state and the forces of extractive capitalism, they have gained the ability to begin undoing colonization in all its dimensions.[31]

We can find examples of reforestation all over the world. The distinction between a true forest and a tree plantation that looks good on paper but in actual fact destroys the local territory is qualitative. The key factor in determining whether a reforestation effort belongs in the first category or the second is whether it is under local control and designed by localized knowledge, as opposed to being under control of the state.

Ethiopia made the headlines in 2019 by breaking world records and planting 353 million trees in a single day. The effort was a government initiative designed to capture media attention and promote ecotourism, though there are many doubts as to its effectiveness. No independent studies are available, but in one typical village, a third of the seedlings were washed away in a flood shortly after planting. Local experts criticized the government for "not considering how the community can be owners of the process" and not thinking about how reforestation cannot be a one day event but needs to be integrated with the economic and ecological needs of the local people who will be taking care of and coexisting with those trees every day.[32] Others pointed out earlier government schemes going back to the '80s that also took a quantitative approach, but that ended in failure because their top-down model was impervious to local knowledge.

Another danger with entrusting the environment to governments can be seen in the ongoing civil war in Tigray, where the Ethiopian and Eritrean governments are systematically destroying the orchards of local people and committing other acts of genocide.[33] Destroying a people's food sovereignty, their capacity for self-sufficiency, is a standard part of counterinsurgency operations, as well as a tendency of capitalism and states in general. Where people are fully self-sufficient, the state cannot exist.[34]

In July 2020 in western Kenya, the Sengwer people of the Embobut Forest and the Ogiek of the Mau Forest were evicted and had their houses burned down. The culprits? The Kenyan Forest Service. Nor was it the first time they had been evicted, not even that year. The Kenyan government employs a particularly colonial form of conservation known as "fortress conservation."[35] Imported directly by, in this case, the British colonizers, fortress conservation is an artifact of the European aristocracy's wars to enclose the commons and to proletarianize the rural poor. It sees humans and nature as antithetical and the purpose of conservation as the expulsion of poorer humans to enable the creation of nature parks, primarily for the entertainment use and aesthetic preferences of the wealthy.

What this practice refuses to see is that humans are a part of our ecosystems and that in the case of the western Kenyan forests, they exist in large part thanks to how the Sengwer and Ogiek peoples interact with them. It also refuses to acknowledge that states are not instruments for protecting the environment, they are machines of exploitation, accumulation, and surveillance. To anyone who understands this, it comes as no surprise that the Kenyan Forest Service is in fact one of the main bodies linked to illegal logging. Critics insist that "governments and conservation organisations continue to fail in applying a human-rights-based approach to conservation, despite numerous international commitments to do so."[36]

In India, a constellation of small farmers and Indigenous peoples have been crucial in protecting the subcontinent's remaining forests.

Similar to Ethiopia, the Indian government has tried to capture headlines with major tree planting events. Under capitalism, after all, it makes more sense to cut down existing forests and to replant new forests, but ecologically, this does not work. In India as in Ethiopia, successful plantations have been "mainly the result of community-led efforts." Government-sponsored tree planting has often resulted in land theft from indigenous Adivasi communities, sparking major protests. Many Adivasi speak out as their traditional rights are "violated for the sake of artificial forests that pay little heed to the local ecology." For example, the government plants native teak trees, but in vast monocrop plantations, rather than as one tree interspersed with many others in a diverse ecosystem that also supports local human populations. Of course, teak plantations make for a valuable export crop, though it's not the local communities who see the profit. The government also favors eucalyptus trees that grow quickly and sequester large amounts of carbon, but the eucalyptus drain the water table and hurt native species, another example of how climate reductionism is actually bad for the planet. To meet its commitments under the Paris Agreement, the Indian government is trying to double its forest cover. But since the technocrats ignore the difference between a forest and a tree plantation, this means fencing off communal, Indigenous lands, and planting monocrops that further harm the earth.[37] Millions of Indigenous people in India are currently threatened with eviction from their lands by the state.[38]

Small farmers have been adapting to the ongoing disaster, employing important innovations even as the government threatens to run them out of existence. While (industrial) agriculture is the second greatest emitter of greenhouse gases on the planet, many small farmers in India are turning to agroforestry or forest gardens, mixing trees with annual crops to provide a more diversified diet, draw carbon out of the atmosphere rather than emitting more of it, and heal the soil (50 percent of which in India is degraded, thanks to the "Green Revolution").[39] Agroforestry also gives small farmers an advantage, because it is a method that favors quality over quantity, allowing more people to sustain themselves with agricultural work—and to do so in a way that is actually good for the earth rather than ecocidal—at a time when the state is trying to dispossess India's remaining peasantry in line with capitalist development models.

They have recently changed agricultural laws in a clear move to favor large, industrial producers and to drive small farmers off the land. In response, hundreds of thousands of small farmers, many of them from Indigenous communities, went on strike at the end of 2020, fighting with police and blockading roads around the capital.[40] According to Pranav Jeevan P, the farmers' movement in India is largely organized according to practices of mutual aid and self-organization, typified by leaderless solidarity and logistical organization around *langars*, free community kitchens that feed the farmers while they maintain their blockades. The blockades quickly encompassed a revolution of daily life, taking on questions of childcare, education, and culture, microcosms of a new world being built rather than a merely political protest for different government policies.[41]

FIGHTING WHERE WE LIVE: FROM CITIES TO HABITATS

The urban-rural divide is a central dynamic of capitalist accumulation and the ecological crisis. There is a differentiated regime of extraction, accumulation, and social control between rural and urban space. Just as rural struggles are rediscovering their potential for blockades and sabotage, urban struggles are learning that they are not limited to protesting and destroying; they can also transform. In order to reclaim cities as habitats, ecological struggles in cities merit special attention.

As a first step, that means keeping cities from killing us. For poor people, urban life is often a death sentence, even as medical infrastructure under capitalism is concentrated in cities.

In the 1970s, New York City officials and business interests began planning to build a trash incinerator at the Brooklyn Navy Yards. The incinerator would have plagued local neighborhoods like Williamsburg with dioxin and other lethal forms of pollution, but Puerto Rican and Hasidic neighbors fought back using a "by any means necessary" approach, taking on the city government, the utility company, and major media that all supported the plan. They definitively blocked the incinerator in 1995.[42]

What should not be elided is that subsequent to this neighborhood victory, Williamsburg and much of the rest of Brooklyn have been aggressively gentrified, with property values going through the roof, and many working-class residents and people of color pushed out in favor of disproportionately white young professionals. In other words, many of those who fought for a cleaner neighborhood were not allowed to stay around to reap the benefits. This kind of story is systematically typical, and a reminder of why the supposedly pragmatic position of partial reform is hopelessly naïve. As long as capitalism remains intact, whatever gains we happen to win by pressuring existing institutions will be enjoyed by economically privileged strata and those who are best able to assimilate to the racist codes and culture of a colonial society.

Another struggle that links environmental concerns with the economic needs of poor urban residents is the defense of public transportation. These can include Critical Mass bicycle protests from San Francisco to São Paulo that oppose car culture and in many cities have led to the creation of bicycle lanes and increased access for poor residents to bicycles and bicycle repair. More than a lifestyle question, cities designed for cars are lethal, especially for residents in denser neighborhoods. Cities that are organized in such a way that workers have to rely on automobiles are simply increasing indebtedness and funneling wages to corporations in two of the wealthiest sectors in the Global North: the automobile and petroleum industries.

The defense of public transportation has also sparked full blown revolts. In Barcelona and the San Francisco Bay Area, mass refusal to pay or public actions to neutralize ticket checkers and open up metro stations for free, whether organized by neighborhood assemblies or anarchist organizations, and sometimes in tandem with labor strikes by transport workers, have temporarily reduced financial strains faced by working class commuters and also generated tremendous pressure on municipal governments against further fare hikes.

In both Brazil and Chile, major insurrections grew out of movements that initially formed in response to fare hikes. Both the 2013 movement in Brazil and the 2019–2021 uprising in Chile counted on a decisive anarchist participation, defeated the proposed fare hikes, and were able to identify a much broader social horizon, expanding to address deeper issues of injustice including police repression, inequality and austerity, and the right to the city.[43]

Urban movements often feel doomed to failure: those who live in a city rarely have any chance to resist changes to their own neighborhoods that are imposed from above. In part, that is because throughout the twentieth and twenty-first centuries, cities have represented the concentration of capital accumulated on a global scale. Legally, houses and other buildings are not places for people to live or carry out their professional activities; they are essentially bank accounts where major interests can safely park the trillions of dollars they have made through currency speculation, private equity marauding, the underpaying of workers, the overcharging of tenants, and the stripping of complex ecosystems to sell for parts. It does not matter who lives there and what they need, or even if these buildings are left empty for decades.

So when we fight for our right to the city, we are going up against capitalism at the point where it is strongest. Furthermore, police departments in major cities today tend to be larger, better financed, and more heavily armed than national armies were a century or two ago. The fact that decentralized urban movements can rise up and force the state to back down (Soweto 1986, Hamburg 1987, Cochabamba 2000, El Alto 2003, Paris 2005 and 2006, Oaxaca City 2006, Athens 2008, Oakland 2009, Tunis and Cairo 2011, Istanbul 2013, São Paulo 2013, Barcelona 2014, Santiago 2019, Minneapolis 2020, Lagos 2020 ...) is extremely significant, and should be a central consideration in any strategy for social change today. However, urban rebellions are frequently excluded from the official conversation. Sadly, cynically, this is a reflection of the disorder and sacrifice they entail—inimical to the culture and class interests of the experts who control the conversation—and a reflection of the difficulties around controlling these movements. Urban rebellions tend to move from single flashpoint issues to ever broader and more revolutionary horizons. Would-be politicians cannot control these movements while they remain active; on the contrary, their main form of influence is a partial ability to demobilize in exchange for short-term reformist gains or, failing that, to encourage internal conflicts in the movements.

By focusing on technological or administrative solutions rather than the decentralized and often combative responses social movements themselves keep offering up, most academics and writers from the Global North fail to tailor their technocratic proposals to the immediate need for survival, dignity, and direct control by people and communities over our own lives. Social justice and decolonization have become buzz words, but most of the people today who are getting paid to make proposals or write about the problem evince a practice that is deeply colonial. Fortunately, we don't need them. Proposals for dignity, survival, and self-organization are popping up like mushrooms after the rain, originating in affected communities themselves.

Cecosesola is a network of cooperatives in Venezuela that took questions of survival and dignity as their point of departure. Cecosesola is currently made up of 50 communitarian organizations located in seven states in the center and west of the country. It began in 1967 as an initiative for cooperative funeral homes. At a moment of familial tragedy that also often implied financial hardship, working class families were being fleeced by private, for-profit funeral homes. The new cooperatives were intended to fulfill this need with dignity and respect, while also lowering costs so as to be affordable.

Following this desire to meet popular needs through cooperative structures, Cecosesola branched out in the mid-70s to provide public transportation. It was a period of massive mobilizations against transport fare hikes, so the network pooled its resources and they acquired a fleet of buses, soon providing the majority of the transportation in the city of Barquisimeto. The experience continued until 1980, when the government seized and destroyed their buses.

Cecosesola also confronted a slow but steady drift into bureaucratization, the same tendency that has turned most other cooperative ventures into little more than alternative businesses. I corresponded with a few members of Cecosesola and they told me how they completely changed their organizational practices to prevent bureaucratization, to remove the positions of supervisors and boards of directors from their legal statutes, and to continue meeting lower class needs in a worsening economic situation, while also expanding their scope to take on ecological considerations. Cecosesola is an important survival mechanism for hundreds of thousands of people amidst Venezuela's major economic crisis. Their cooperative markets are the principal source of food and supplies for some 100,000 families, they link farmers with urban residents, build up communitarian production of pasta, coffee, cleaning supplies, and other goods, help finance harvests or the purchase of vehicles, and provide healthcare to 230,000 people a year. Throughout our conversation, the members I spoke with gave the most attention to questions of process and relationship-building, a clear difference from the dominant institutions' focus on quantitative results.

Food insecurity is also a problem in the Global North, especially in racialized neighborhoods. Cooperation Jackson, in Jackson, Mississippi, arose in 2014 from the experiences of the Black liberation movement and mutual aid efforts in the wake of Hurricane Katrina. The organization establishes cooperatives to build a circular economy with an emphasis on gaining access to land within the impoverished, majority Black city, growing healthy food that can be prepared and distributed by other cooperatives, and collecting waste and producing compost and fertilizer by yet other cooperatives in a "reinforcing value chain" that expands to cover a growing body of vital needs. They also set up a manufacturing cooperative to "promote the arts and science of digital fabrication" and fulfill the goal of "owning and controlling the means of production."[44] This project, in turn, is part of the plan for building the Ewing Street Eco-Village, a permaculture-based communal housing project for cooperative participants, and a springboard for yet more cooperative enterprises.

Cooperation Jackson follows a strategic plan that seeks to "[build] autonomous power outside the state structure" while also "engaging in electoral politics on a limited scale" largely as a means to protect themselves from "the power of the state," which, "as we learned through our own struggles," can present a grave "peril" to movements.[45] Their solidarity economy of worker cooperatives and "informal affinity-based neighbourhood bartering networks" is "grounded on the principles of social solidarity, mutual aid, reciprocity and generosity."[46] Following an ideal of ecosocialism, Cooperation Jackson prioritizes survival in its many dimensions, from gaining employment and economic self-sufficiency for members, increasing access to healthy food, reducing dependence on hydrocarbons, and transforming their neighborhoods to improve the quality of the air and water. In these efforts they have received support from local trade unions and progressive churches.

Cooperation Jackson has achieved considerable successes in a politically and economically hostile environment. Member Kali Akuno identifies two prime challenges or setbacks their model has faced, which parallel the experiences of related initiatives from the Mondragón cooperative complex of the Basque country to Plan Pueblo a Pueblo which we looked at in an earlier section. The first is the tendency of cooperatives to focus on profit margins and reintegrate with capitalist economies. Following this trend (which in a matter of decades has made the Mondragón complex indistinguishable from the rest of the capitalist landscape), one of the original Cooperation Jackson cooperatives broke away from the group, as they were not "prepared to make the required sacrifices necessary to promote ecosocialism through the development of a non-exploitative and regenerative system at the community level." The second is all the work that remains to be done "getting people to understand the severity of the crisis *and* our collective ability to do something about it."[47]

This in turn relates to the difficulty of convincing the government "to implement the policy proposals" that are needed, that will only materialize, Akuno argues, if "consciously driven by the people."[48] An ally and veteran of the movement for racial justice, Chokwe Lumumba, won the mayoral elections for the city of Jackson in 2013, and it seemed the grassroots would receive official support for their projects and urban transformation plans. But Lumumba died prematurely after just a few months in office and subsequent politicians have not been supportive. This turn of events underscores the importance of Cooperation Jackson's emphasis on building autonomous power and not waiting on electoral cycles.

Given the statist differentiation between city and countryside, control over food and artificial scarcity have been frequent methods for controlling urban populations. As the counterpoint, urban gardening is often a way for people to assert neighborhood control in the face of real estate developers and racist urban planning experts, to improve quality of life, and increase community bonds.

In North Lawndale, Chicago, there are some 20 garden spaces. Many, like the MLK garden on 16th Street, have a long history of significance in local antiracist movements. These gardens, which often start with neighbors directly taking over plots, organize skillshares, hold community events, improve air quality, and distribute food throughout a neighborhood that experiences real food scarcity. Several gardens donate medicinal herbs to a historic Black pharmacy and support formerly incarcerated people. In Lawndale, Little Village, and other neighborhoods, the urban gardens are connected to a network of associations arising from Black, Puerto Rican, Indigenous, and migrant communities. One such organization, Reclaiming Our Roots, "is an initiative of Southwest Chicago that reclaims public spaces for healing and gathering using indigenous knowledge. Reclaiming Our Roots creates opportunities for youth to engage with a community garden while connecting and caring for the land."[49]

The "Special Period," in which Cuba faced sudden food shortages when they lost access to petroleum due to the collapse of the Soviet Union and a US embargo, shows how quickly urban populations can transition in order to feed themselves, in this case through the explosion of urban and suburban gardens that ordinary people planted and tended to using hand tools and small-scale techniques. Although the Cuban state encouraged the phenomenon, it was decentralized in nature, and some who lived through that moment tell me the main thing the government did was to get out of the way and allow communities to self-organize. Indeed, there is little evidence to suggest a government would even be capable of organizing such a profound transformation in such a short time frame.

Urban reforestation is also an important line of action to consider. Whereas government-led programs tends to focus on ornamental species and to actively avoid creating a free commons of edibles, neighborhood-initiated efforts can carry out planting in a way that simultaneously removes carbon, decreases pollution and urban warming, autonomously provides food security, and turns cities into habitats that can be healthy for humans and other species.

Small City Forest Gardens, Somewhere in the US

I spoke with one person involved in anticapitalist initiatives in the United States, including an informal effort that has resulted in the planting of over 10,000 fruit-bearing, native trees and shrubs in a couple of neighboring small cities. Because their circles have been targeted multiple times by police and FBI for their anticapitalist and ecological politics, they have wished to use a pseudonym and keep their location anonymous.

Victor explained how they take advantage of a government subsidized program meant to help commercial farmers, using it to buy native, fruit-bearing trees in bulk and distribute them in different communities:

> It's worth noting that the entire agricultural industry is structured around state subsidies. What the USDA [US Department of Agriculture] funds or guarantees thrives, most else withers. The state decides, among growers, which farms are made to live and which are left to die, and this apparatus and resource monopoly is something that we have to slowly break while also extracting resources from it.
>
> This is why it is so decisive that Black farmers have been historically excluded from USDA loan programs. In bad crop years, they have to pay the real cost of their wages, while white farmers stay insulated, and thus Black farmers have lost a huge portion of the land they held 100 years ago. So it's important that the USDA recognizes "five agroforestry practices." These are magic words and mean that these practices open the sluice gates for cash and subsidies, in the same way that other recognized farming practices do. So farmers get paid to plant in windbreaks, and they get the trees cheap too. They mostly

go for big hardwoods that they plant as investments for logging, but it creates a leverage point for us to pull resources out of this system and create our own reservoirs for cooperative growing and future plant breeding and distribution.

Our species list includes pawpaw, American plum, hazelnut, American persimmon, elderberry, aronia, and chokecherry. These are all made available via subsidies but are rarely intended to be planted with enough density for human food or medicinal production and instead are commonly planted to meet one of the other 5 recognized agroforestry practices such as riparian plantings. We also order hardy berries like currants and gooseberries in bulk from commercial suppliers, and select trees like serviceberry from friendly nurseries. We solicit donations from these friendly nurseries as well—which is a way that people who have already made the move back-to-the-land can express an organized complicity—and this year we've received mulberry, blueberry, blackberry, hardy fig, and chestnut as donations.

We have slowly spread out and included more neighborhoods in town as they've requested participation, but we organize it each year within core neighborhoods that we do other autonomous organizing in. I think that the project has been successful in producing collective spirit in certain neighborhoods and not others. We give people the choice to take trees to plant on their own or to receive help, or to use trees to plan neighborhood plantings. All of the neighborhoods have held at least one of these collective plantings. Two of the neighborhoods did seem to really draw inspiration from planting in trees together. One is a neighborhood with housing projects and the other is a neighborhood of mostly Latin American immigrants who are used to growing their own food and mixing residential and agricultural spaces.

As we've spread the plantings to more neighborhoods, we've also been confronted with class differences—and even relatively minor difference in income—between areas. These have posed problems in terms of resource distribution that can't always be easily solved from an autonomous perspective, especially regarding trees which are in short supply or most apt for particular places.

So far, mostly only the berry bushes have produced fruit, but once the fruit trees mature it will be a game changer in terms of creating a commons and many working class people will have free access to a huge amount of fresh fruit and nuts. One of the challenges we face

> is redeveloping familiarity with some of the native plants that were never appropriated by the colonizers for commercial food production. Some of these non-commercial fruit are delicious raw, like pawpaw, but cannot be sold because they are too fragile to ship to stores, and thus perfect for neighborhood consumption. Others, however, are best enjoyed with processing, like the so-called chokecherry. First Nations peoples have many ways to process fruits like these, and so we have researched and discussed the appropriate ways to draw on these methods.

Such an initiative can "scale up" by filling out, with just a few active people in any region or small city adopting the model. Spread throughout an entire country, this practice can plant an atmospherically significant number of trees. Unlike with state-led initiatives, there is a solid guarantee that the trees will be taken care of and will actually survive to maturity, accelerating processes of community building and commoning that will be necessary for lower class people to confront the ecological crisis with resilience. It improves quality of life for lower class people and creates stronger relationships to place that can serve as an effective pole for resistance against the notion of private property wielded by real estate developers and speculators, enabling a huge part of the map—small cities or less dense large cities—to play a significant role in carbon removal. This is a preemptive mutiny against the imminent plans of the technocrats to turn a huge chunk of rural space into sacrifice zones for carbon removals so that life in the cities can continue with minimal transformation.

Grassroots, autonomous efforts to vitalize the urban habitat can also be a source of powerful resistance against municipal attempts to restructure a neighborhood in the interests of capitalist profit and top-down urban planning.

Vallcarca is a peripheral neighborhood in Barcelona, formerly an independent village, that has become the site of a particularly complex resistance bringing together *assemblees llibertàries* (anarchist assemblies), neighborhood entities, and self-organized communal spaces. In 2002, city planners slated Vallcarca as a sacrifice zone for a motorway expansion that was to destroy the entire old town center. Speculators bought up, emptied out, and demolished properties throughout the center of the neighborhood. Anarchists and others began squatting many of these properties to prevent their demolition, while neighborhood associations protested the devastation. More recently, neighbors of all kinds have come together to occupy the vacant lots left behind by the demolitions and plant urban gardens filled with vegetable plots, activity spaces for important traditions like *petanca* and *calçotades*, and hundreds of fruit and nut trees. The network of gardens brings neighbors together and encourages many more people to participate in the defense of the neighborhood, as well as providing a space for cultural events, socialization, and play; taking a first step towards food autonomy; and improving the quality of the neighborhood in terms of health, carbon footprint, beauty, and enjoyment, for the benefit of the lower class people who live there rather than for the higher income demographics the real estate speculators will attempt to bring in.

With weekly blockades shutting down traffic, occupations of evicted and demolished spaces, the unmediated creation of plazas and gardens, and other forms of protest and direct action, the neighbors forced a change to the official urban planning document for the neighborhood in 2018,[50] and far more importantly, they created real changes on the ground and took important steps towards recovering the power to make decisions and organize their own lives.

In the nearby neighborhood of La Salut, a group of neighbors have taken over a vacant lot to create a garden with a global focus. The garden itself is organized in large part by people who have had to traverse continents and cross borders fleeing the effects of a mercenary capitalism, giving birth to one initiative within the garden space, La Enredadera, a *semillero diaspórico* or nursery garden for people of the diaspora. In the words of one participant:

> Many of the people who have migrated to Barcelona have been displaced by climatic and agricultural pressures, direct and indirect, although only loosely categorized. Under a wider ecosystemic lens, our migration can be considered as an adaptive climate strategy, of survival and resistance, with benefits for our territories of origin as well our territories of arrival. ...
>
> The climate migrant leaves behind profound relations to their territory and generally passes through a painful mourning while struggling to start again from scratch, but they also hold onto to valuable knowledge about modes of life in transition, resilience, and adaptability.

In sum, La Enredadera responds to the question, "how do we root the concept of global climate justice at the local scale?" using "the tools of documentary films, cultural geography, and agroecological design" in collaboration with the rest of the garden project.

Though from the beginning, the gardeners took a more conciliatory approach towards the city government, emphasizing dialogue, in the end the progressive administration evicted them from their original plot without availing a second location they had promised. In order to acquire a second home and continue their project, the group had to occupy a high value developing property to force the government to honor their promises. It might be noted that this was a left-wing administration happy to take photo ops to file away in their resume under the headings of "multiculturalism" or "green urbanism."

La Enredadera makes clear the connections between migration, colonialism, the ecological crisis, and urban resistance and occupation. As migration and ecological crisis become increasingly enmeshed, initiatives focusing on winning housing for and by undocumented people also deserve a central focus. Once again, this is a field in which anticapitalist movements using direct action lead the way.

Throughout Catalunya, there are numerous initiatives by undocumented people and their accomplices to squat abandoned warehouses to provide housing for large groups. Numerous squatted social centers linked to the anarchist movement also house migrants. And there are several initiatives for self-employment by undocumented people, from making and selling handicraft bags to cooking for events, that form a part of anticapitalist networks and movements.

Such initiatives take on a greater scale in Gran Canaria, an island in the West African archipelago occupied by Spain, and with the highest rates of poverty in the whole Spanish state. The Canary Islands are also one of the main points of entry into the European Union for African migrants. The Anarchist Federation of Gran Canaria (FAGC) has occupied several housing blocks and set up entire autonomous neighborhoods, organizing for basic survival among lower class people, both documented and undocumented. Working with the offshoot organization, the Sindicato de Inquilinos (Tenants' Union of Gran Canaria), they stop hundreds of evictions every year, and they have occupied numerous properties to provide "socialized" housing, meaning free and self-organized, liberated from both exploitation by landlords and humiliation by government bureaucrats. At the time of this writing, a thousand people live in housing socialized by these two organizations, and thousands more retain access to capitalist housing thanks to the blocked evictions. Some of these socialized housing units are collective or communal, from a squatted hotel to "La Esperanza," a self-organized community with 76 families, more than 200 people. Two other projects, "Las Masías" and "El Refugio II," provide housing to around 100 and 150 migrants, respectively, who face legal persecution, while "El Nido," a squatted school, serves as a shelter for women fleeing domestic violence. The FAGC has also squatted four large agricultural plots for autonomous food production on the island, and they provide a consultation service for workers facing labor conflicts. Most of those who use this service are informal workers, including sex workers who have experienced sexist and transphobic violence.

Their methods have been so effective that government social workers on the island have often advised unemployed and low-income people looking for housing, food assistance, or other resources to "go to the anarchists."[51] Already, these activities surpass the scale, and certainly the quality, of government alleviation measures. The main limitation to spreading even further is not any internal defect of this kind of organization. On the contrary, we can identify two external factors. The first is state repression. One of the most active organizers with the FAGC was arrested and tortured by police, and then in a typical move in such cases, falsely accused by police of assaulting them. He is currently facing trial.

The other limitation is that such a practice requires organizing from below, which means a loss of control for professional activists. Similar initiatives organized by and for those in need of housing, using protest, blockades, and occupations, with migrant women frequently taking a leading role, are active throughout the Spanish state, including the Plataforma de Afectadas por la Hipoteca (PAH), which has been featured in numerous articles in English-language media. However, at key moments, middle class activists have proven their ability to flex within the housing movement and prevent an evolution towards stronger, more generalized tactics like rent strikes that would put into question the property relations at the foundation of capitalism and achieve the ideal that *houses belong to those who live in them*. The tendency of reformist activism towards a single-issue focus also keeps such movements from spreading out of control, which they need to do to actually and effectively challenge globally interlinked systems of exploitation and oppression.

In Greece, the other main entry point for migrants into Fortress Europe, there have also been inspiring initiatives providing housing and the means of survival. Tens of thousands of refugees from Syria, Afghanistan, and dozens of other countries, working closely with the anarchist and autonomous movements, squatted entire hotels and other abandoned buildings in Athens, Thessaloniki, and other cities in order to attain dignified housing.[52] The squats are self-organized by their inhabitants, and many of them reject NGO workers as well as direct representatives of the state. They also run cafeterias, language classes, clinics, childcare services, and other activities. Some of the squats have waiting lists, with thousands of people hoping for a place. In contrast, the camps run by the government and managed by NGOs are literal concentration camps, with people living in unhygienic conditions in tents or barracks, with no services, eating disgusting food sometimes infested with maggots, freezing to death in the winter, and suffering countless abuses.[53]

The progressive Syriza government had little sympathy for radical movements that put the party's complicity with business as usual in sharp relief. They evicted several migrant squats and tried to repress the movement more generally.[54]

In the face of the 2017 wave of repression, the movements rose to the challenge, demonstrating that principles like solidarity, mutual aid, and self-organization are not just feel-good ideas, and that in moments of crisis they scale up effectively, in contrast to the NGOs and the progressive political parties, who very literally *became the problem*, and in contrast to the chorus of Global North academics who enjoy the old game of infantilizing anarchistic political action but in fact have little to show when it comes to the housing and border crises. Anarchists and others occupied ruling party headquarters, worker-run factory Vio.Me opened "a warehouse for the collection, storage and transportation of basic items like clothes, sanitary items and baby food that had been gathered by solidarity collectives from all over Greece and Europe, prior to shipment to the Eidomeni border to be handed out to refugees," thousands of people invited the most vulnerable among the refugees into their homes, and side by side, migrants and Greek radicals opened the next generation of squats.[55]

As usually happens, the failure of the institutional Left to either satisfy social demands or to restore *order* opened the way for conservative parties to go ever further to the right. When the New Democracy government came back into power in 2019, they declared an open war on self-organizing migrants and the anarchist movement more broadly, evicting numerous squats with heavily armed police and kicking undocumented people out to sleep on the streets or sending them into the system of state-run gulags.

In a sense, poor people are not allowed to be at home anywhere, papers or not, and across the world the struggle for a right to the city is raging. In poorer countries in the Global South, the outcome of that fight can often determine whether people have to migrate in the first place, or whether they can build the collective power to make a home.

Abahlali baseMjondolo, whose name in Zulu means "people who stay in shacks," is "a radically democratic, grassroots and entirely non-professionalised movement of shack dwellers" in Durban, Pinetown, Pietermaritzburg, and other parts of Western Cape, South Africa. The movement began in 2005, arising out of a practice of road blockades and land occupations carried out by poor shack dwellers in Durban. They reject participation "in party politics or any NGO style professionalization or individualization of struggle."

> Since the 2005 road blockade the movement's membership has grown from the entire population of the 6000 strong Kennedy Road settlement to the point where 13 entire settlements have voted to collectively affiliate to Abahlali and govern themselves autonomously from state politics. There are also a further 23 branches in other settlements that are not collectively affiliated to Abahlali but which do allow independent political activity.
>
> This means that around 30,000 people have a direct and formal affiliation with the movement but many more have been inspired by it.
>
> … The movement now also works with street traders and has a further 3 branches of street traders, all of which are in the city of Pinetown. It also has members in two areas in Pinetown in which people live in poor quality houses rather than shacks and who joined because they became familiar with the movement as their communities are adjacent to Abahlali shack settlements. The movement is multi-ethnic, multi-racial and multi-national and operates on the principle that everyone living in a settlement is from that settlement and has full rights to participate in the political life of that settlement irrespective of their origins.[56]

Abahlali has fought for access to housing and land, education and child-care, water and sanitation, electrification, they have "held quarterly all-night music, poetry and drama evenings; run a 16 team football league; provided HIV/AIDS care;" and they have also built relationships of international solidarity "in support of shack dwellers in Zimbabwe and Haiti." One of their main goals is "to win popular control over decision making that affects poor communities." Doing so has directly pitted them against "authoritarianism from government, business and some left NGOs with vanguardist delusions." Due to their activities and also their independence from the dominant institutions, their members and activities have been repeatedly attacked by police, sometimes with tacit support from poverty-focused NGOs.

Urban gardens are common in the settlements of South Africa, as they are in informal cities throughout the Global South in which one billion people live. Due to the framework of development, such gardens face a particular set of problems linked to ongoing colonial dynamics. Many residents of these impoverished, self-constructed, and largely self-regulating urban areas are recent evictees or transplants from the countryside, where a combination of capitalist and governmental pressures make life ever less tenable and desirable. However, urban agriculture is not exactly a continuation of rural agriculture, given the huge differences in the necessary techniques, as well as the world-crushing pressure that identifies small-scale agriculture with rural places and therefore as something backwards that needs to be left behind. In the Global North as in the Global South, urban agriculture is practiced as a survival strategy when food insecurity grows, and one that often persists and spreads when people realize its many social and environmental benefits.

"[M]any national and local governments in sub-Saharan Africa have resisted [urban agriculture] and urban farmers often face harassment by officials and the police." People trying to feed themselves, improve their quality of life, and adapt traditional food cultures to their new surroundings are confronted with militarized discourses of hygiene that signal what forms of life should be emulated and what forms of life are an obstacle or a threat. "Urban agriculture presents an impression to local elites and to international observers that the city remains 'unmodern', 'uncivilized', 'uncontrolled', and 'under-developed' … Within this rhetoric, it is possible to detect a form of anti-Africanism."[57]

The need to transform cities into environments where food is abundant, ending their existence as barren sites of accumulation that despoil the countryside and where the poorer classes are especially vulnerable to malnutrition, and where there are few niches for other species, runs into a wall. Yet the limitations are not material. One of the primary obstacles is the very notion of progress, which is inextricably colonial and Eurocentric. Drawing inspiration from Frantz Fanon,[58] Faranak Miraftab draws attention to the colonial imagination of urban planners and challenges:

> the assumption that every plan and policy must insist on modernization. This mental decolonization requires recognizing how the ideal of the Western city has been deployed historically in the colonial era and is now being deployed in the neoliberal era to advance a certain paradigm of development and capital accumulation. A collective of developers, planners, architects, and politicians and a powerful industry of marketing and image-making have promoted the Western city as an object of desire.[59]

Given more space or a two-volume book, I would showcase autonomous technologies developed directly by urban residents, like the rainwater filtration and greywater bioremediation developed in the anarchist social center, Casa de Lagartixa Preta in Santo André, Brazil, or bee habitats and rooftop gardens cultivated from Den Haag to Detroit. Instead I have chosen to focus not on the particular technologies that will accompany the transformation of our cities, but on contests over who controls the space. I have no doubt we can deploy all the amazing innovations that have been developed across the world the moment our neighborhoods belong to us.

The point is, the technologies to transform cities into healthy habitats already exist. We are not lacking inventors, we are lacking control over our own lives and vital spaces. Until we can directly organize and transform our neighborhoods to meet our own needs, and break the monopolies that control the world's resources—including intellectual property—new technologies will be of two varieties: bootlegged, autonomous ingenuities developed *in situ* that make the most out of scarce materials; or engineered technologies developed by professionals, well-meaning or otherwise, that will only increase global inequalities.

4
Versatile Strategies

A THOUSAND WORLDS STRUGGLING TO BE BORN: ECOSYSTEMS OF REVOLT

The struggles and initiatives described in the last chapter constitute a revolutionary wave that can be found in nearly every country across the world. The examples I've provided are just a tiny sample of an extensive web of obstruction, sabotage, demolition, healing, cultivation, creation, learning, and communication that represents the best hope for our planet. It is the only force currently in existence that meets all the following criteria: a structural independence from the bodies responsible for ecocide and colonizing capitalism; a capacity to force the state to back down in key conflicts; access to the locally specific knowledge necessary for real and intelligent responses to the unfolding climate catastrophe; a tendency to break through barriers and create an increasingly global consciousness that centers an awareness of the intersection of all forms of oppression and all the unfolding crises; access to traditions of organization and ecosocial relation that open the possibility for a world without capitalism, without ecocide.

Please do not mistake my glowing review for optimism. This is still a battle that pits David against Goliath, and if we were to approach the ecological crisis as though it were a wager in a casino—as the economists do, for example—then our money would be more wisely placed backing the forces of the apocalypse. However, if we wager our lives—they're on the line already, whether we've realized it yet or not—this motley network of underdogs is our best hope. All the other proposals for responding to the ecological crisis are some variation of the strategy in which David becomes Goliath's shield bearer with the hope that, over time, Goliath will begin using his spear for good.

What are the limitations of this revolutionary wave?

The primary external limitation, as we have seen, is the counterinsurgency being waged against us, from moments of hard repression—all the people we have lost, all the people currently sitting in prison for their struggles—to the soft repression and invisibilization that mainstream environmental groups, media, and experts participate in, willingly or unwillingly. If at least some of those who are currently throwing their energies into redeeming Goliath were to shift their resources to supporting this revolutionary wave—which would also mean losing their considerable institutional privileges—then our chances would be considerably improved. There is a lack of a revolutionary imaginary and a lack of consciousness that these different movements constitute seeds for future worlds. Basically, this means withdrawing our remaining faith in the dominant institutions and believing more in ourselves and the future we are trying to move towards. This is a limitation that is already being overcome within and between these movements themselves, and this book represents one small effort in that direction. It is, in other words, neither fatal nor insurmountable.

There is a go-to mischaracterization that is used to dismiss the revolutionary potential of this wave, and it is very much an expression of the dominant institutions' need to monopolize society-wide organization and problem-solving. It is the aspersion that these movements have no solution to offer that would be feasible on a relevant scale. To name one iteration, Holly Jean Buck characterizes the phenomenon Naomi Klein names as "Blockadia" as "reactive."[1] I won't go into whether this dismissal is the result of a misreading by Buck or of the limited view Klein utilizes to present Blockadia as more palatable (begging the question, palatable to whom). I will offer the observation, though, that experts are trained to silence their subject of study, so it does seem both systematic and symptomatic that, in looking at such a rich phenomenon, ranging from the resistance at Standing Rock to Hambach Forest, one would see something "reactive." In any case, the broader, more global, less respectable view of the resistance that I have tried to present makes it clear that we are dealing with something intelligent, creative, strategic, proactive, and with a whole hell of a lot of proposals that will not be silenced.

The movements and projects that form this global web are marked by their heterogeneity, heterodoxy, and plain ornery refusal to be easily categorized. I do not think this *many-headed hydra* of resistance should be named; after all, a being with a thousand heads would come up with a thousand different names for itself.[2] However, I do want to name common characteristics in the most flexible way possible, to encourage what I see as strengths and to aid more people in transforming their own activities so as to be able to connect, rhizomatically, with this greater whole. The following characteristics are not bounded containers that can govern inclusion or exclusion in a delimited phenomenon; rather they are tensions that vibrate throughout the entire web.

Territoriality

A relationship with the specific local territory constitutes a main source of power for these struggles and projects. We develop our practices and histories in dialogue with the territory such that "the environment" is neither inert surroundings nor a neutral field on which to impose an ideology that is the same from place to place.

Rooting struggles in a specific territory walks a line between two forms of isolation. In nearly every struggle, there will be people who limit themselves to their territory, who do not look for common ground with other struggles or seek inspiration from their own experiences that could have a wider, perhaps global, significance. And on the other hand, there are those who are alienated from any territory even as they participate in "local activism." They draw ideological lines for solidarity; either they restrict solidarity to their own small sect, or they read the values of their sect into all of those with whom they would like to solidarize. Such people are a part of the mix, and this is a complication in territorial struggles, but also a form of openness, presenting the possibility to weave in a wider body of people.

Ecocentrism

While many of those who constitute this revolutionary wave prioritize human needs, we tend to reject the pretension that human needs can sustainably contradict, outrank, or detach themselves from ecological needs, and at one level or another, we challenge or reject definitions of humanity stemming from the European Enlightenment and human/nature dichotomies.

Survival

We articulate our activity in relation to situations that directly affect us and we center this struggle as a question of survival, our own and that of other people and forms of life we care about. Having a voice, therefore, does not come down to expertise or institutional legitimacy, but to being affected and personally engaging with the problem and its solutions.

Lawlessness

To a greater or lesser degree, these projects enter into conflict with established legal regimes. They may actively seek the subversion and destruction of existing governments, they may claim traditional and Indigenous systems of law (that paradigmatically have nothing in common with punitive or property-based law originating in states), or they may seek as much as possible to pass unnoticed or mold themselves to existing legal regimes, but they will always value the needs of their community and the needs of the earth more than the authority of the government or the ostensible sanctity of its law.

Communal being

Individualized or atomized views of human beings are eroded in favor of practices that emphasize and revitalize relationships between people (sometimes including relationships that break down the division between humans and other forms of life). There is always an element of struggle against the alienations imposed by states and capitalism, and a tendency to practice mutual aid and solidarity. This means that this web of resistance is fundamentally creative: of different social relations, different subjectivities, and emancipatory infrastructures, practices, and cultures.

Heterogeneity

As mentioned, this "movement of movements," to use the Zapatista's terms, is extremely heterogeneous. This does not mean simply diverse, but that it is constantly producing differences and that it will not submit to ideological or cultural unity. This salient feature makes any technocratic proposal for solving the crisis extremely ignorant, which is probably why the technocrats tend to ignore or selectively silence movements that already exist when designing solutions. It should be noted that this heterogeneity is not just a cultural preference of the network, it is an inalienable feature of the network's territorialized nature. This is why ideologies or named traditions of struggle that are structurally capable of assuming diversity rather than unity of practice—such as *Zapatismo* or anarchism—would never be able to absorb all the iterations of this revolutionary wave. The only reason these traditions of struggle are tolerated and encouraged throughout much of the network is precisely because they have no ideological need to convert others to their way of thinking or to achieve theoretical unity.

Intersectionality

The movements that participate in this wave tend to break down single-issue containers and instead recognize the interconnectedness of different forms of oppression and, therefore, solidarity. This intersectionality allows us to recognize one another even though we come from very different places and lack uniform identifiers. The process of recognition, it should be noted, is conflictual rather than pacifying—people often fight to get recognized on their own terms, a struggle that is not made easier by the ways in which we have been differently socialized within oppressive systems—meaning the self-definition of struggles is crucial to the possibility of solidarity across the network: people are implicitly trusted to define their own oppressions and lead their own struggles. This is another death blow to any pretensions of imposing uniform solutions.

Anticolonialism

All of these initiatives and movements exist in contradiction to the project of development, which is the most active manifestation of colonialism in the age of the International Monetary Fund, the World Trade Organization, the United Nations, and all the attendant NGOs. Beyond this common negation, there is a great deal of distance between fully anticolonial movements, movements that identify capitalism as the enemy without exploring colonization as a historical and ongoing process integral to the globalization of capitalism, and even movements that use the language of development in a bid to access resources or legitimacy. Even among the former currents, there are very different experiences of colonialism across the world, but the heterogeneity of the resistance means those differences do not have to present a problem. I would argue that, despite the broad differences in language and scope, these movements' practices open up possibilities for complementarity, and that an expansive anticolonial consciousness is a priority for increasing their potency.

In a use of these terms that is far from universal, I think it is useful to make a distinction between "decolonial" and "anticolonial." The latest buzzword, "decolonial" is now frequently used in academic and activist texts that make no mention of the restoration of Indigenous lands and don't even have the decency to so much as hint at the possibility of abolishing settler states that owe their existence to colonialism, like the United States or Canada. What exactly is decolonization supposed to mean, if the fruits and vehicles of colonialism are grandfathered in and accepted as eternal? The distinction I would make is between movements that seek reconciliation and disarmament, and those that seek to destroy forces that have been accepted as universal. These latter movements hold out hope for a victory that will undo some of the defeats of the past 500 years (or 2,000 years, or longer, depending on the territory we are looking at).

Autonomy

People who constitute this international network may be actively trying to subvert and destroy the state, or they may be looking for some breathing room from state repression in which to carry out their activities; some may even support an alternative government that might reduce the degree of repression. We may believe that the contest with the state is central to our struggle, or that the state will disappear if people gain some form of economic self-sufficiency. We may reject any contact with the state, or we may try to win access to government resources. Whatever the case, a general practice of autonomy is what allows us to work together and to form cohesive networks of resistance. Autonomy means we write our own rules, we make our own decisions independently of oversight by any party or government, whatever the provenance of our resources we make the final decisions about how those resources are used, and we practice self-organization and avoid the centralization of the movement.

If one element of these movements maintains a relationship with a political party or a government, they take care not to let that relationship condition their activity in the movement or convert them into a lever by which the government and party can exert influence over the movement; if they fail to uphold either of these minimums, it is widely seen as a breach of solidarity by the rest of the movement. Without autonomy, it is impossible to create a movement of movements, a world in which a thousand worlds can flourish.

These last two characteristics especially—anticolonialism and autonomy—are the sites of crucial strategic debates. I am very much a partisan in these debates, and I firmly believe that spreading the practices of anticolonialism and autonomy, more or less as I have described them but certainly without any doctrinaire consensus, is vital to improving the revolutionary potential of our interconnected movements.

The framework of autonomy enjoys a special advantage. It is often the default political compass in peasant struggles and Indigenous struggles[3] as well as urban squatting and countercultural movements.[4] The 2006 Oaxaca rebellion that created a large pocket of stateless self-organization for five months throughout a large part of the southern Mexican state drew heavily on Indigenous traditions for their organizational structures. In the framework of federated assemblies, labor organizers and peasants, Indigenous peoples and *mestizos*, peasants and city-dwellers, anarchists and Marxists could all work together.

It has become increasingly common for famous writers and green academics to give a nod to social movements by claiming that they are a necessary adjunct to political change. This allows them to acknowledge that the dominant political institutions have a terrible track record when it comes to solving social and ecological problems, without being so radical as to identify those political institutions as one of the sources of the problem. However, they diplomatically avoid any historical analysis of how the relationship between states and social movements always ends up.

In her bestselling book in favor of the Green New Deal, *On Fire*, Naomi Klein reproduces a speech she gave to the UK Labor Party in 2017, while Jeremy Corbyn was leader. In this speech, she urges for a strong relationship between the Party and the social movements, claiming that social movements "have a huge amount to gain from engaging with electoral politics" though she is remarkably short on details.[5] We didn't have to wait long to see how poorly that wager panned out. Corbyn's own party sabotaged the progressive agenda, ensuring that the only electoral options for the public would be the right and the center.

Though I am an anarchist, I have never held a dogmatic position against voting. What is most important is what you do with all the other thousands of hours in a year. The reason I now caution people against engaging with electoral politics is that I have seen the consequences, a clear pattern from the Corbyn campaign in the UK to the Sanders campaign in the US to the austerity referendum in Greece to the independence referendum in Catalunya to the MAS victory in Bolivia to the constitutional referendum in Chile. Aside from the immense energies and resources wasted in these campaigns to sway a vote rather than to build infrastructure for communities in struggle, there is a clear psychological result: every single time, when they lost, and even more so if they won, resilient movements that supported these electoral campaigns with a justification of urgency or gradual change became jaded, burnt out, and demobilized in the aftermath. It is as though the symbolic act of voting carries a very real psychological weight, as though we were depositing our hopes in a machine that will inevitably disappoint us. Political parties are unscrupulous in how they will cannibalize social movements and suck them dry. Even in countries where political engagement is most easy to justify on grounds of mere survival, for example where the old government has direct ties to the military and death squads, alliance with a more progressive government constitutes a dead end, with initial improvements giving way, the pendulum swinging back to old, bloody habits, in a matter of years. The truce is at best a temporary affair. Context is more important than dogma, and people's survival is paramount. Those who live in a situation where their very survival is threatened shouldn't have to worry about the uninformed disapproval of those for whom the danger is less urgent, but they also shouldn't have any illusions about the inevitable trajectory of state power. This is why building autonomous power and understanding that all the institutions of colonial society endanger us is so important.

FALSE PRAGMATISMS: STRATEGIES OF DESPERATION

Lake Lleu Lleu, located in the Biobío region of southern South America, is one of the cleanest lakes on the entire continent.[6] Located in the heart of Wallmapu, the Mapuche country occupied by the Chilean and Argentinian states since the end of the nineteenth century, Lleu Lleu stays clean thanks to the concerted action of the Mapuche communities that surround it. When *winka* or colonizer companies start building cabins to attract tourists and in their wake the promise of jet skis, vehicle traffic, and the sheen of oil and suntan lotion atop the water, Mapuche warriors burn those cabins down. When the *carabineros*, Chilean military police, build outposts in the region to protect business ventures from arson or to guard the extensive forestry plantations from Mapuche land reclamation actions, they get burned down too. When mining companies come into the region to prospect, their surveying stakes get removed and their vehicles get destroyed. When the state tries to crack down on this resistance, the communities block all the highways in the region and organize combative protests in the nearby city of Concepción or the Chilean capital, Santiago.[7]

This ongoing victory is a microcosm in which we can tease out the deficiencies of the major proposals for solving the ecological crisis coming from the Left. Staying within the bounds of the law or setting the limit at respectable and symbolic civil disobedience is simply unrealistic. In any territory where inhabitants are not collecting the wages of whiteness, there is no clean water without arson, blockades, and clandestine organization. The Mapuche struggle around Lleu Lleu also belies the idea of decolonization as just a polite word tacked on to the end of a policy proposal that relies on the very institutions of coloniality to solve the problem. In real practice, it means physically and forcefully taking land back, incurring tremendous risks, and entering into conflict with all the institutions of dominant society.

Their successes are also a warning against climate reductionism. Within even a moderately technocratic gaze focused on statistics of carbon reduction, Mapuche communities actually become the enemies of the climate and further repressive action against them can be justified with an environmental logic. As we have seen, a part of the Mapuche struggle involves destroying commercial forestry plantations, the very plantations that, from a global, technocratic view, count as vital carbon sinks and a major part of the Chilean government's strategy to become a celebrated "carbon neutral" country. Only with an intensely local perspective can we see that the plantations are not forests and they are actually destroying the land and depleting the water. Once we abandon the technocrat's office and adopt that local perspective, other issues become as important or more important than statistics on atmospheric carbon: defending a sacred lake, restoring specifically native forests, and fighting for food sovereignty. None of these goals are prioritized by the climate movement in the Global North, but in fact they represent a path of struggle that is far more able to address the ecological crisis, including global warming, than a reductionist approach.

The civil disobedience movement Extinction Rebellion suffers from climate reductionism in multiple ways. By making almost no reference to the long and diverse history of ecological struggles it is cutting itself off from the collective learning experiences that infuse our histories of resistance.

As I have pointed out elsewhere, historical amnesia is a key feature of nonviolent movements, especially in the last twenty years.[8] By not engaging with the lived histories of struggles and the lessons they present, nonviolent movements shelter themselves from the tactical and strategic evolution that tends to disfavor a dogmatic nonviolence. They also avoid deep analysis of the social situation they face, which protects them from having to study, for example, the repressive mechanisms of the state, or how nonviolence in social movements is a key objective of counterinsurgency strategies.

In the case of Extinction Rebellion, they substitute movement histories with statistics promising the success of nonviolent civil disobedience. The statistics, incidentally, come from a study that compares a handpicked list of ostensible pacifist victories with a database of bloody military conflicts, intentionally leaves out movements that demonstrate the effectiveness of a diversity of tactics, fails to define victory, frequently misrepresents its own results, and was authored in part by a former employee of the US State Department and Defense Department.[9] Needless to say, the statistics were spread widely by corporate media.

The political naïveté of Extinction Rebellion extends from their methods to their goals: namely, the idea of pressuring governments to declare "climate emergencies" and to support the Paris Agreement. The declaration of a climate emergency is meaningless, in policy terms. It is a public relations Christmas for governments on the right and the left, a way for them to clean up their image with another hollow declaration. As for the Paris Agreement, we have seen how inadequate it is. Subsequent action goals like "money strikes" also tend to be weak on their own, as well as exclusive to those who have money; and the creation of Citizens' Assemblies with actual decision-making power would require major constitutional reforms, an outcome that hasn't been achieved by peaceful movements. But again, the proposal is easy to co-opt, as governments love consultative bodies with no real power as a way to institutionalize and neutralize demands.

Looking at the movement with the very urgency they call for, it seems like Extinction Rebellion is using up an immense amount of resources and the energies of tens of thousands of young people in order to lobby for a few buckets on board the Titanic.

But it is worse than a question of inefficiency. Extinction Rebellion's insistence on a single issue focus—they have been frequently criticized for apathy to environmental racism and the marginalization of racialized people and struggles[10]—fails monumentally to address the central question of colonialism and opens the door to ecofascism.

The possibility for a renewed alliance between ecology and white supremacy was established at a critical juncture of the environmental movement in the last century. In the 1960s, environmental consciousness was expanding, manifesting in the works of Murray Bookchin, Rachel Carson, and Vine Deloria Jr. During this period, Indigenous struggles achieved global attention, and movements against imperialist wars, including support for the rights of refugees, also blossomed. People could have chosen an integral consciousness that saw the connections between these different struggles. But one sector gravitated towards the defense of an anti-human environment at the exclusion of all other issues, perhaps lured by the donations of the old conservationist aristocracy. Thanks to their unsolidaristic exclusions, we can talk about environmentalism as a white supremacist phenomenon, of most interest to progressive whites but equally useful to fascists of the "Blood and Soil" variety who mimic the German Nazi Party's cooptation of early twentieth century back-to-the-land movements.[11]

Not only has this alliance sullied a large part of the animal rights movement and made inroads into the climate movement, it has leapt to the international stage, as when the Austrian Green Party formed a ruling coalition in 2020 with the highly xenophobic ÖVP or Austrian "Volks" Party, who had previously ruled with the literal neo-Nazis of the Freedom Party.

It remains to be seen whether Extinction Rebellion will have staying power, or whether it will disappear like other astroturfed movements before it. Whatever the case, the strategy of lobbying the government for emergency measures will remain dominant as long as there exists a capitalist media to redirect all social conflicts into dialogue with the institutions of power.

The Green New Deal is the main proposal on the Left, and in a social democratic framework it makes a lot of sense. It directs massive public spending to pay for a total infrastructure overhaul, completing the transition to renewable energy, while also increasing social spending to ameliorate the inequalities that have provoked growing crises in nearly all the world's democracies. At the same time, government-subsidized production may be enough to trigger a new industrial expansion and leave behind the crisis of accumulation global capitalism has been dragging in its wake since 2007, and in some senses since the 1970s. In a stroke, the Green New Deal could resolve the three interconnected crises that threaten the state: the ecological crisis, democracy's crisis of legitimacy, and capitalism's crisis of accumulation.[12]

In truth, the GND should be catechism for the most ardent defenders of the current system, toppling as it is, but it seems that nearly everyone is one long step behind events. Conservatives, locked in their perpetual existential crisis, are too threatened by the updating of their privileges that they are further destabilizing the system they are trying to protect, as always.[13] Centrists are trying to preserve a middle ground where there is none. And so it has fallen on progressives, many of them avowed anti-capitalists, to become capitalism's last line of defense.

Economists have largely acknowledged that the current model of permanent deficit spending by governments is in fact feasible, with governments always spending beyond their budgets.[14] However, it is only possible to avoid major defaults and economic collapse if governments continue to service interest payments, which they can only do if their economies keep growing, giving them a higher GDP and higher incomes year after year. The massive expenditures of the Green New Deal require an increase in deficit spending which in turn requires a constant increase in the overall volume of the economy. Any Green New Deal would mean giving up on any hopes of degrowth in favor of the dubious proposition that infinite economic growth can in fact be healthy and sustainable on a finite planet. Some have tried to revive this pipe dream by conjuring up a slanted portrayal of the information economy, suggesting that its processes are creative, affective, and ethereal; therefore not dependent on material extraction. The Bitcoin boom—with appropriately named Bitcoin miners now using more energy than many mid-sized countries—should put to rest any doubts that capitalism is always fundamentally extractive. All value chains eventually come back to killing ecosystems and selling the corpses.

Despite promises of spending for social justice programs, the GND remains climate reductionist. It will increase the amount of destruction caused by so-called green energies by several orders of magnitude, add new devastating infrastructures for carbon capture and storage, spur wars and coups for copper, lithium, and cobalt, and it will create a multi-trillion-dollar market for land—carbon sinks, sites for wind and photovoltaic production, biofuels—that will be the death knell for small-farmers, peasants, and Indigenous communities the world over.

By richly incentivizing carbon sinks, world governments and NGOs are laying the groundwork for a massive land grab that is set to match the scale of the invasions and colonization of the sixteenth to eighteenth centuries. Expanding the market for carbon—whether through tax credits or major government subsidies—will be a huge boon to the logging industry, to energy companies, and to agribusiness as remaining agricultural lands will have to shift towards more intensive, infrastructure-heavy, market-oriented production.

It is easy to see why Wall Street and the luminaries of Big Business tend to be tepidly in favor of the Green New Deal or similar proposals.[15] The GND means a massive amount of government funding and tax credits, and companies like Microsoft, Tesla, Google, and forward-thinking oil companies like BP are the best positioned to capture that money. The main point of reticence is that the GND would increase government intervention and control over the economy, just as the original New Deal constituted a transfer of "control over world liquidity ... from London and Wall Street to Washington" in the words of Giovanni Arrighi.[16]

Nearly all the projects we looked at in Chapter 3 are short on resources. All of them can use more support. But none of them are calling for billions of dollars of infusions in government money, precisely because they know that would destroy them, monetizing the cracks in the system where they have been able to thrive, bringing in major corporations eager to profit off human and ecosystem services, and in the Global South increasing the funding for paramilitary death squads far beyond what the drug and logging industries were ever able to provide.

A reference to Franklin D. Roosevelt's New Deal that pulled the US out of the Great Depression, the GND set itself up for an illuminating, and accurate, historical comparison. The original New Deal brought to bear immense government resources to stimulate the economy and create employment. It is often associated with the beginning of Keynesian economics, governments running deficits to actively intervene in the economy in coordination with a central bank. Though the New Deal is lionized by progressives, it presents a curious allusion for the environmental movement, given how it was exemplified by the construction of megaprojects like hydroelectric dams that displaced tens of thousands of small farmers and destroyed entire river valleys, or more than 1 million kilometers of roads and highways that scarred the land and constituted an infrastructural handout to the automobile industry that would come to dominate the architecture of American life, as well as becoming a principal source of air pollution. True, the New Deal also included reforestation and soil conservation efforts in response to the Dust Bowl. However, government aid was designed not to restore the health of the land but to continue commercial agriculture. The Dust Bowl only ended when rainfall patterns returned to normal in the region, and while soil conservation improved, other ecocidal practices continued, as evidenced by ongoing desertification today as well as the dead zone in the Mississippi River delta.

The New Deal has a sheen of social justice due to a few programs that challenged oppressive inequalities. The Federal Writers' Project, for example, paid writers to record oral histories of certain populations, most famously documenting the memories of thousands of elderly Black people, the survivors of the last generation to live through slavery. While there is a clear value in such projects of historical memory, the Federal Writers' Project manifested and reinforced the very inequalities that were being addressed, as typically occurs when governments throw money at a problem of social justice: most of the writers getting government checks to survive the extreme misery of the Depression and record the memories of slavery's last survivors were not Black but southern whites, many of them from families who had profited off of slavery and off of the sharecropping that followed it.[17]

Even Naomi Klein, in her book in favor of the GND, acknowledges that the original New Deal was deployed because it "appeared at the time to be the only way to hold back a full-scale revolution."[18] It provided the hand-outs and the funding to weaken or institutionalize social movements, getting them under control, and within the framework of well paid, institutionalized labor unions and full employment underwritten by the government, it replaced the revolutionary internationalism of the workers' movement—a movement that launched mutinies throughout World War I and countless general strikes, insurrections, and revolutions at the war's end—with the narrowed horizons of patriotism. The New Deal set the stage for the weakness of revolutionary movements after World War II, for the nationalism and conservatism of the Cold War and all the assaults on workers' rights and social justice that entailed.

Historically, social democracy has always planted the seed of its own destruction.

That's exactly what the Green New Deal does. It prevents a real solution, it kills off our best chance for a better world, in order to save capitalism, to kick the can down the road, or more realistically, into the laps of the people who will bear the brunt of this new industrial onslaught. The Green New Deal is colonialism as usual.

Asad Rehman calls it "green colonialism" and points out that the green energy transition of the Green New Deal is reliant on the same economic structures that have caused the problem. He also underscores the danger of the mining boom a green transition would require. "Resource extraction is responsible for 50 per cent of global emissions" as well as a great deal of violence against poor communities from the Democratic Republic of Congo to Chile, he says.

> At the heart of our economic system fuelled by the City of London is a belief that the UK and other rich countries are entitled to a greater share of the world's finite resources irrespective of who we impoverish in doing so, or the destruction we cause.
>
> This green colonialism will be delivered by the very same entrenched economic interests, who have willingly sacrificed both people and the climate in the pursuit of profit. But this time, the mining giants and dirty energy companies will be waving the flag of climate emergency to justify the same deathly business model.[19]

By trying to capture existing channels for economic action on a global scale, the GND will inevitably plug into the institutional pathways and Eurocentric framework of development, a thinking and practice that has been roundly criticized as racist. As Rajni Kothari put it, "where colonialism left off, development took over."[20]

Massive funding for green energy and industrial-scaled carbon sinks is already producing an outsized impact on the Global South. In Mozambique, "consistent with other African contexts," the biofuels industry, expanding as a direct result of European governments' renewable energy targets, "legitimizes and greenwashes acts of land-grabbing, to the detriment of some of the most marginalized in society … causing entire communities to be displaced, leading to food insecurity, resource deprivation, social polarization, and political instability" and "constitut[ing] vast spiritual and cultural loss."[21]

Good intentions are about as widespread as bad results. The pervasive racism of the development paradigm, of the existing institutional pathways for directing global interventions, is too systematic for this blunt instrument to be corrected with a more liberal application of decolonization rhetoric.

Some authors and organizers do take these tendencies into account, and they offer proposals for a Green New Deal that would make sure money directly reached those most in need, that it would be used to empower communities and overcome the legacies of racism and colonialism without funding the hyperexploitative markets I have warned about. The problem with these proposals is that they read like a wish list for an ideal world. They do not come with examples of government programs that have ever achieved similar results in the past.[22] If the Green New Deal were actually carried out in such a way, it would strip the very people who make up the government of their wealth and power. Even stable democracies like the United States and United Kingdom would see a military coup before those who hold power would allow a policy one half as radical to be legislated. Nowhere is there a government with more than a tiny handful of representatives willing to propose such policies.

Another proposed response to the ecological crisis is eco-Leninism or "war communism" for the climate crisis. Some iterations of this proposal focus on the method: creating a vanguard cadre organization to lead the climate movement; and others focus on the finality: strong governments that will ruthlessly use their regulatory and punitive powers to get their houses in order.

In terms of substance, this proposal is little more than tough talk. It does pose a danger though, because if the past is any guide—and it usually is—authoritarian and vanguardist formations can grow quickly when the dominant institutions are facing collapse. Their destiny is usually to make some kind of alliance with those in power, either to betray, institutionalize, and repress revolutionary movements, or to create a new system with many components of the old one and all the oppressive baggage they drag with them.

The comparison to "war communism" can be attributed to Andreas Malm.[23] Like the Green New Deal, it seems a poorly thought-out historical allusion. The USSR was a fully ecocidal regime. As a revolutionary experiment it left the most interesting questions unanswered. How much might we have learned from urban soviets and factory councils creating direct relationships with rural councils of peasants and food producers during the Russian Revolution? Lenin prohibited this direct self-organization by the proletariat between November and December of 1917.[24] How much might we have learned from Alexander Chayanov, the author of *Peasant Farm Organization* and a proponent of small-scale cooperative farming as opposed to large-scale state-run farms? He was executed on Stalin's orders in October 1937. As for war communism, that was nothing but a bald euphemism used to describe the Red Army's bloody plunder of the countryside to keep alive the two primary instruments of Bolshevik power during the civil war: the Army itself, and the urban Party bureaucrats. War communism was substantially identical to what any army will do when it doesn't have supply lines to a productive and stable hinterland. It corresponded not to any vision of social transformation, but to the interests and power struggles of the moment. How are we to interpret this as a proposal that states today might carry out? What state will use its regulatory powers so ruthlessly, not in its own interests but against its own economic elite? And what major states even have a vanguard party remotely close to being able to grow over the next few years, take over, and implement this war communism against polluters?

The proposal is completely out of touch with the social movements that actually exist today, the movements that comprise the only force for change not beholden to the economic and political interests destroying the planet. The movements behind nearly all the major revolts since the collapse of the USSR share several emergent characteristics: multitudinous actions largely at the margins of all formal organizations; a skepticism towards dialogue with existing institutions and instead a tendency to disrupt their managerial dominance; a revolutionary organizational framework that is specifically anti-vanguardist, that seeks to create spaces of dialogue across social movements rather than controlling such movements, and that breaks with the statist practice of subsuming revolution to a military contest; and an emphasis on assembly-based decision-making structures, frequently traditional and Indigenous in origin, as more legitimate and more liberatory than state structures.[25] Even historically Leninist movements from Venezuela to Kurdistan have been shifting to more autonomous frameworks. The social movements that exist today are not building the kinds of power that would be useful to the project of taking over the state, and that is very good news.

Since the post-war demassification of the proletarian and the growing strength of anticolonial, antiracist, and antipatriarchal struggles have multiplied the lines of struggle, homogenizing organizations like parties have not been able to express the aspirations of social movements. They have tended towards insularity and social conservatism, or they have taken advantage of sudden booms to make inspiring promises, forestall their inevitable collapse, and rush into the halls of power, institutionalizing as much of the social movement as they can and gaining access to government funding and more opaque forms of financing as the only way to stay alive.

I think the major reason such unrealistic proposals—all coming down to trusting the government to solve things—have been flourishing is desperation. We feel weak, millions of people are dying, millions of species are threatened with extinction right now and every year emissions only increase, the point of no return might already have slipped by, our movements are not as strong as they need to be, and so it seems that the best hope is for someone to wave a magic wand and make things better. We certainly don't have any magic wands, but from down here it seems like the government might.

The protections governments offer are often illusory. They have a perfectly abysmal track record for honoring their emissions targets, their biodiversity targets, their treaties. Ashley Dawson's book *Extreme Cities* is full of examples of how major cities from Miami to New York, faced with the existential threat that they will soon be under water, have wasted decades with bureaucratic regulations that are little more than evasions.[26] So how will government regulation be transformed into a machine efficient enough to hang our lives on? This seems to be the crucial strategic question for those who pin their hopes on the state. How do you make the state function differently to how it has always functioned? Where are the models from the past that show states work effectively to protect their people—all their people—and their environments, without resorting to some kind of resource imperialism?

Occasionally, we can find vague suggestions that social movements will force governments to stay the course. So if social movements are the key component for resolving the climate crisis, why don't all the books that advocate a Green New Deal or a robust regulatory state give social movements a voice, and make social movements the protagonists of their story, sharing their specific experiences, methodologies, and proposals? Why don't they present us with histories of social movements and governments achieving fruitful collaborations? They do not, because they cannot. Because the more they propose governmental solutions to our problems, the more distanced and deafened they are to the movements, and because the history of our movements shows that the relationship with governments, on balance, is an antagonistic one.

Governments are machines designed to allow the upper classes to exert pressure on the rest of society. They do not work the other way around; they are about as reversible as a telescope. They limit the horizons of possibility, rather than expanding them.

Nor is it sensible to place our faith in progressive or revolutionary governments, because even these remain part of an interlocking global system they cannot simply wish away. To fund social programs and import food, Hugo Chávez in Venezuela had to encourage oil extraction to generate foreign revenue, leading to violent conflicts with Indigenous communities whose lands are degraded by oil production. Nonetheless, cheerleaders in the Global North are easy to fool. In Tiquipaya, Bolivia, in 2010, when Evo Morales hosted the World People's Conference on Climate Change and the Rights of Mother Earth, his administration merely had to hide all the sawmills on the road leading to the conference site with billboards, confident none of the international attendants would speak with locals or inquire after ongoing deforestation. A few years later, he mobilized his union supporters to attack Indigenous marchers with dynamite as they protested the construction of a major highway through their territory, exposing the western Amazon to a development boom. It was the kind of megaproject the earlier right-wing government would never have been able to get away with. In fact it benefited from the superficial anti-imperialism spread by the populist Left, that reduces everything to puppeteering by the US government, ignoring how capitalism actually implants itself in every territory it touches. While the devastation was carried out by Bolivian companies and Brazilian capital, its opponents were denounced as agents of the CIA, culminating in the aforementioned attack as well as an antiterrorism campaign against local anarchists.[27] Silvia Rivera Cusicanqui characterizes the Morales government as "crav[ing] centralized power" and "putting an Indian face to corporate capitalism."[28]

Across the political spectrum, governmental environmentalism endangers us, it does not make us safer. As environmental "inputs" and "outputs" are increasingly monetized, subjected to rational measurement, we draw ever closer to the dystopia of a technocratic control of the environment which will also translate into more technocratic control of humanity. (Neither is this an overly paranoid vision: "Smart City" plans systematically bond green proposals with increased, AI-based surveillance, and MI5 was recently caught funding research on agricultural drones that also doubled for rural surveillance.)[29]

Reflecting the class interests and the worldview of the technocrats themselves, all of these proposals enact power as a lever that operates on inert Others. To them, the territory is a map, and theirs is the hand that holds the pencil that will redraw it. Deep down, they can never trust the intelligence of the territory (nor locate themselves within it). They cannot surrender themselves to the dialogue, the dance, with a specific territory, nor meld into the reciprocal relationship that is *the earth healing itself.* Modern day missionaries, they fail to accept that they are not needed to save anyone. And that is why they remain a part of the problem.

To quote Erahsto Felício again, of Teia dos Povos, "Those who have not yet territorialized and organized their community cannot lead a fight for land and territory!"

Since I am an anarchist, it would be easy for many of those whose proposals I am criticizing to chalk up my rejection of the state to a merely ideological position. For me, though, anarchism does not mean converting everyone to my way of thinking, it is a methodology for building a world in which a thousand worlds can fit. The proposal that heterogeneous social movements require autonomy from the state in order to work together would be rather limited if only fellow anarchists saw the benefits of such autonomy. So I spoke with someone who, like me, prioritizes ecological struggle and rejects the colonial project, but who has a very different political background.

In Chapter 2, Adrianna Quena described the interrelated problems of expertise and whiteness. We also learned about her experiences in the *chavista* movement in Venezuela, where she was a part of numerous initiatives that sought government resources or that collaborated in some way with the government. Often, this was a way to win some "breathing room" after the brutal right-wing governments of the past and the ongoing violence of the US- and Colombian-backed economic elite that still holds a great deal of power in Venezuela (a stranglehold, it might be noted, that Chavez was unable to break and that in some cases has intensified). Calculations of urgency make sense in this context. "For us, climate collapse already began some years ago."

Even though we differ on the proposition of a revolutionary government, it is heartening that we come to many of the same conclusions regarding the tendencies of governmental action and the need for social movements to cultivate their autonomy. A key figure that took on a central importance in our conversations was the technocratic gaze, how it constitutes a continuation of colonialism, maintains its hold even on revolutionary governments, and is fatal to any lasting solution for the ecological crisis.

> At one extreme, we have the "philanthropy" of foreign NGOs. For example, I can briefly describe the case in which the French oil company Total, as part of its *greenwashing* policies, had the great idea of installing solar panels in the Amacuro Delta in Venezuela, so that some Warao communities could have access to electricity. In this case, a concrete need such as being able to charge mobile phones, provide lighting for the schools, run a television for the families gathered around to watch kung fu movies, run a refrigerator, is satisfied through an expertise in renewable energy, but going over the heads of the communitarian Indigenous logics of authority, decision-making, and management of resources, and then afterwards never taking care of the maintenance and leaving the Warao territory in an absolute mess: full of trashed panels, "renewable" in name only, and with various intracommunity conflicts. Another failed attempt by the modern, technocratic paratrooper method [or *paracaidismo*, a rather visual metaphor for the experts who parachute out of an airplane with no idea where they'll land].
>
> On the other end we have the governmental institution, that sends a technician to try to resolve a local agricultural problem, for example, with a solution straight out of an agronomy manual that is nothing but propaganda for the poorly named "Green Revolution." They worry about output, as though it were an isolated affair. They are obsessed with the quantitative because that is what they can write up in their spreadsheet. They are concerned with their technological packet (seeds, agrotoxins, machinery), sold to them by a provider, almost always a foreign company, who is already entered into the ledgers. They don't have the capacity to dialogue with that which already exists, they don't see the potential of an agro-ecosystem beyond what is written in their manual from the Netherlands or Canada. Their recipe requires bringing in almost everything from the outside, sweeping away that which already exists, forgetting local knowledge, the comprehension of the relations that exist there, ignore the spiritual dimension, of course, and convert everything into a neutral substrate so as to begin from zero, like in the laboratory. This subject—whitened, modern,

and technified—does not listen, and with his design he degrades. And if, as in this case, he is a representative of the state, vested with the authority of the very institutionality that measures your access to basic rights, it's even more difficult.

It is always a matter of whiteness and specialization.

There is this myth that the territory can always be controlled and domesticated. Maybe in territories where domestication and the imperial logic have been established for thousands of years (like large parts of Europe), this can make sense, but in many of the places that are today known as the Americas, the history is different: the dialogue with the territory and with all the beings that inhabit it has not been one of domestication by control and subjugation, but rather of coexistence and symbiosis.

If we root the topic of quality and sustainability in more political dimensions, connected to the climate, we can see how in Venezuela, the merely quantitative and homogenizing logic of the technocracy obstructed the development of popular power (which is power that is territorialized and from below), the diversification of the economy, the redesigning of education, in the most opportune moment for all those things, which was during the government of Hugo Chavez.

This transformed the self-perception people had of their own agency and generated the construction of a new local political class that—because it can speak the language of the institutions, it can adapt its mentality to these quantitative approximations, a reductionism of goals, objectives, number of beneficiaries, zeros and commas—accumulated privileges and ignored problems, often the most important ones, instead solving less pressing problems that were easier to express in their language.

The government is almost entirely constituted by the so-called *mestiza* middle class, or more accurately white passing and urban, and they do what they can, but with the interpretative limitations of not knowing the territory and with their ambitions of completing the process of becoming white.

The technocracy is impatient, it needs quick results to fill in the numbers in the spreadsheet so it can say whether a project was implemented or not, to guarantee votes and so on. This idea of a mixed public-communal model always tended more towards an inefficient statism and less towards the communal. It gradually disempowered the peasant organizations and the families, and this opened the way for state management that enabled networks of corruption controlled by old and new capitalists, but in the hands of the state rather than in the historical rich families of the area.

The greater the autonomy and the popular control that was achieved, in terms of organization and evaluation defined from within the territory, the greater the sustainability they have demonstrated in the long term, even under the immense pressures of an international blockade and a pandemic.

ECOLOGICAL REVOLUTION: THE BEST STRATEGY IN SUCCESS OR FAILURE

Transforming the solutions we project also means transforming our understanding of the kind of methods that will get us there.

"Today all belong to the ruling class who look at their own lives from above." The anarchist essay *23 Theses Concerning Revolt* elaborates this idea of an embodied strategy:

> All military strategy is to impose an ideal plan on the map that represents reality. … If we do not intend to make a military campaign, we must refuse to see the revolution as something organized according to a unified plan, as if it were a game of Risk. We are not looking down from above, giving orders. We are here, in the midst of a beautiful chaos that our enemies always try to organize. We will be stronger than ever if we learn to triumph in this chaos, to move in the network of our own relationships, to communicate horizontally or circularly, to use only what really is ours and to influence others, to understand that not everyone is going to act as we act; that is the beauty of rebellion, and our effectiveness in it does not lie in making the whole world equal, but in devising the best way to relate in a complementary way to those who are different and follow different paths.[30]

The paradigm shift in our concept of revolution dovetails with the paradigm shift in our relationship with the planet and its inhabitants. We need to perceive "an ecosystem of revolt" in which, instead of trying to control what everyone is doing, promote the right ideas and suppress the wrong ideas, we understand our niche and create relationships of reciprocity with those around us, making us all stronger and healthier rather than making us all the same. It is such an ecosystem that is most able to combine collective knowledge and local, situated knowledge, allowing the decentralization of intelligence rather than continued reliance on the experts and leaders who have brought us into this crisis.

A part of this approach means accepting that we will never convince everyone. Others will continue to fight in legal channels seeking limited reform or government funding. There can still be instances of synergy, even with those who remain in such a diametrically opposed paradigm. On the one hand, we can take advantage of resources intended for other uses, such as the people in the small town in the US planting thousands of trees subsidized by the state. In other cases, we might get to know sympathetic bureaucrats or academics who are able to pass resources to our projects and support our struggles, as with a couple of academics who supported this book, or the radical academic from Borneo who uses their research to support Indigenous communities in struggle.

Each alliance should be evaluated on whether it helps us or hinders us from building the kind of decentralized, communal power that we need. Rejecting technocratic unity or Party dogmatism in our movements does not mean embracing an acritical relativism in which all ideas are equally valid. It is not an ecosystem of revolt if there is no revolt. One of the most important tasks today, which we must spread as energetically as we can, is to develop revolutionary horizons with as many people and communities as possible, precisely because the state, with all its appendages, has spent so many decades extinguishing even the thought of revolution.

Revolutionary movements come with several key advantages. Only by fighting for revolutionary transformations that present an existential threat to the existing order—for example, a world without rich people, without police, without governors, without corporations—have we ever achieved important concessions. All of the labor laws, paid vacations, limited working days, rent controls, public healthcare, unemployment benefits, and safety regulations that are being chipped away year by year are the vestiges of reforms won by revolutionary movements of anarchists and communists in the first part of the twentieth century, or since then by the organizations arising from those movements and institutionalized as a way to pacify them.

In other words, when we fight for revolution, we win even when we lose.

We cannot, however, view revolutionary movements as simply a bargaining chip. As long as the power to organize our lives is alienated from us and concentrated in the institutions that rule us, any improvement we are granted will be steadily worn away starting the moment we no longer present a threat.

Revolutionary movements taking direct action to "destroy what destroys us" raise consciousness and force all of society to talk about problems the media and government had caused us to ignore. When we meet in the streets in the heat of rebellion, we place all our isolated problems in common and realize how much we share.

Whether you are looking at fracking rigs in Pennsylvania or wind parks in Oaxaca, big businesses and the governments that support them operate in a cloak of secrecy, isolating people and getting them to sign away their lands one by one. When people fight back, such as when they built a barricade to stop armed detachments from erecting wind turbines in Bíi Hioxo, Oaxaca, they "created space to question these projects" and even though that particular wind project was constructed thanks to a complex counterinsurgency campaign that deployed all kinds of mercenaries, from paid assassins and neighborhood snitches to marketing strategists, NGO workers, and academics, battle lines were drawn that continue to facilitate resistance in the rest of the region, while the people who nominally lost this conflict were able to negotiate better conditions.[31]

In the United States, for over forty years, the dominant narrative was that the Civil Rights movement of the '50s and '60s had ended systemic racism. It was only when multiracial, Black-led crowds fought police and smashed wealthy businesses in Oakland after the police murder of Oscar Grant in 2009, and when they did the same on a much larger scale after the police murder of Mike Brown in Ferguson, Missouri, in 2014, that white supremacy became the subject of public conversation. This was thanks to a completely acephalous movement that came alive through many other revolts of varying size across the country, and that blended spaces of conversation, spaces of healing, spaces of occupation and transformation, and spaces of counterattack, vengeance, and wealth redistribution. Significantly, most of the mainstream media that kept systemic racism going strong by serving as the propaganda arm for the War on Drugs and avoiding serious reporting on any of the facets of white supremacy, by 2014, were forced to acknowledge that racism was still a problem.

And as uncomfortable as it is for those who would turn the movement into a lucrative NGO that refuses to support its prisoners, or into a Democratic voting machine, in 2020 that movement manifested again in a huge way and forced all of society, at least for several months, to consider a very radical idea that just a few years earlier would have been inconceivable at that scale: abolishing the police entirely. And the most eloquent opening gesture to that debate was to manifest the idea of abolishing the police by burning the Third Precinct police station of Minneapolis down to the ground. Similar to the Arab Spring of 2011, the revolt in Minneapolis sent shock waves across the world, advancing conversations about racism and spreading tactics of collective self-defense against the police in the UK, France, Brazil, and beyond.

The ability of movements with revolutionary horizons to raise consciousness and shift the limits of the debate is beyond question, but it is also fully wrapped up with another important concern: shifting the balance of power. In the US especially, the police are a central institution. Abolishing them means abolishing the US in its entirety as a capitalist settler state. It was unavoidable that anyone unable to conceive of such a revolutionary horizon would gradually replace abolishing the police with defunding the police, and from there many of them would move on to support increased police funding in order to pay for the cameras or improved sensitivity training that would supposedly put an end to hundreds of years of history.

If we do not have the power to abolish such a huge, ugly machine as the US and replace it with something better, healthier, very few of us have the power to imagine abolishing its component parts. Likewise, if we do not have the power to take over our territory, feed ourselves through ecocentric relations, and abolish ecocidal industries, few of us will have the power to imagine any solution that does not involve the government coming in and waving a magic wand.

Having a revolutionary horizon requires us to build power; a revolutionary kind of power that corresponds to our own needs and ways of living. Fortunately, revolutionary movements are uniquely suited to building decentralized power. In contrast, reformist movements only strengthen centralized power, both by solidifying the authorities' role as arbiter and by funneling massive amounts of resources away from social movements for the purpose of lobbying.

Envisioning revolutionary horizons and building power out of our own activity and the collective satisfaction of our own needs and desires—the kind of power that cannot prop up a new oppressive system—inevitably reveals all the intersecting lines of oppression and resistance. This is the terrain in which we see how climate change, industrial pollution, patriarchy, borders, white supremacy, mass extinction, accumulation by dispossession, and colonialism are inextricable. By definition, this kind of revolution means changing the paradigm. In a society in which even environmentalism has become a tool for ravaging the Earth and its inhabitants, smashing the dominant paradigm is a necessity we can no longer elide.

That is why we need to fight for a revolution, and try our hardest to win. Putting an end to a system that by its very nature produces misery would be the best future for all of us. If we fight in a way that is consistent with the world we want to defend, then every step we take, even if we don't make it all the way, leaves the world a better place and could make the difference between life and death for a person, a community, an ecosystem, that right now is under threat. Recognizing that we live only thanks to a beautiful web of relationships with countless other beings, nurturing those relationships, is the best possible way to live. It is a revolution that transforms our present, honors our past, and offers the best hope for ending an apocalypse that has been going on for too long, moving towards a future where we can all, finally, begin to heal.

5
A Truly Different Future

ECOLOGICAL IMAGINATIONS: PLOTTING A TOTAL TRANSFORMATION OF HUMAN EXISTENCE

A huge amount of resources have been spent to make it impossible for us to imagine a world free of capitalism, free of hierarchy, free of the institutions that originated in colonialism. As such, the only kind of imaginary that is articulated and practiced in dominant society is that of the technocratic engineer drafting blueprints onto a passive territory.

One of the most potent weapons against such interventionism is situated imagining, looking at the world around us, tracing the relations we have and could have, listening to their needs, and giving those needs free rein to develop, to see what directions they pull us in.

If you did not take your eyes off the page after reading that, do so. Give yourself a moment to understand what I'm really getting at. "The world around us" is not an abstract figure. It means the ground underneath your feet. It means the organisms that provide the air you breathe, and the machines that poison it. It means the food in your pantry or refrigerator, the machines that gave it its present form. The land this food comes from, and perhaps other land, much closer at hand, from which it could come, but does not. And the why of it all.

Have you looked at the world around you yet? Do so. Stop reading.

The stories of resistance in Chapter 3 all constitute proposals for resistance with immediate relevance. But they also suggest a horizon. What do we need to do to get to that horizon? Who are the people, what are the institutions, standing in our way? What are the traumas and calcified social relations that make it excruciating to take even the first steps, and what will we have to do to heal those traumas and transform those power relations?

Because we cannot accomplish what we cannot imagine, I will attempt to describe it not as a blueprint, but as an imagined future. The image is global, but not universal. I will focus on one territory that I know intimately—the one where I have lived for most of my adult life—in order to suggest a whole world made up of similarly empowered territories, each unique in how the crisis manifested there and how people came together to solve it.

This section is also an invitation to a much larger book. It is a phantom book of unwritten imaginaries, all in dialogue. Each one of you is writing your own chapter, from your own territory, as you read on.

Catalan countries

A few decades in the future, global capitalism has finally been dismantled, but we are still living with many of its effects. Global sea rise has already proven catastrophic in many parts of the world. Cities built on porous bedrock, like Miami, have suffered massive infrastructure failure and are being abandoned, while many Pacific island nations have been evacuated.[1] The Catalan countries, despite their extensive coastline, have so far avoided the worst impacts of sea level rise thanks to their mountainous topography and steep gradient at the coasts. On the other hand, a renewed relationship with the sea constitutes one of the primary responses to the ecological crisis. An end to commercial fishing trawlers, cruise ships, and oil and chemical tankers, as well as the phasing out of international cargo ships, have given the Mediterranean a much-needed chance to heal, all but eliminating chemical and noise pollution, and halting the stripping of the sea floor.

Kelp and sea grass forests have rebounded, spreading from small coastal preserves off the coast of Catalunya and Eivissa to replenish most of the littoral ecosystem. Much of this was spontaneous regrowth. With the destruction of dragnets, the sea floor was protected, and sea urchin overpopulation, which took its toll on the forests, was brought into balance as crab populations increased thanks to the decline in pollution. Also, the human tradition of consuming sea urchin, marginalized during the regime of capitalist food production, has experienced a major revival. In the traditional food culture of the Catalan countries—Rosselló, Catalunya, the Balearic Isles, and València—seafood was an important food source for coastal populations. The capitalist simplification of local food culture was carried out without any consideration for environmental impact. But now, food culture has been popularized again, and is practiced with intimate knowledge of the ecosystemic balance. When people go to the beach, they often come back with a small basket of sea urchins to add to their supper. This provides a first example of a broader trend:

work and leisure, production and play, are self-guided and indistinguishable. Gone is the alienation that separates work from life. Gone too is the more recent technological trend that had extended the logic of productivity to every moment of waking.

In addition to creating the conditions for the regrowth of kelp and sea grass forests, human communities in the Catalan countries have carefully carried out several programs in which scientists have worked together with coastal communities in a relationship in which the latter have the final say. One was the reintroduction of the Eurasian otter—next to humans the major natural predator of sea urchins—at specific coastal sites where it was determined they would have a viable habitat. Another was a collaboration between marine biologists and diving enthusiasts to plant new kelp and sea grass forests in locations where they were most likely to take, and that could also serve as bridges or stepping stones between remaining forests to increase the possible corridors for species movement and migration during a time of increasing instability.

These practices have preserved marine biodiversity and saved several threatened species from extinction, improved the food security of coastal communities, established buffers against coastal erosion during the super storms that will likely continue for the next couple of centuries, and through the exponential increase in marine forests they constitute one of the main carbon sinks in the region, taking carbon out of the atmosphere as we slowly return to preindustrial greenhouse gas levels.

Another change in the human relationship to the sea involves seafaring. In the most intense years of the transformation, as effective state power waned, remaining fuel reserves were largely apportioned to international cargo ships bringing emergency supplies to vulnerable parts of the world. This largely consisted of two things: food shipments to countries in the Global South that had been forced into a dependence on imports; and machine parts so that poorer regions could attain self-sufficiency producing food, medicine, and all other vital supplies.

This global transfer of resources that allowed poor populations to leave the exploitative relationship of forced dependence was possible thanks to labor organization in the ports and factories. Despite decades of neoliberalism, labor organizing in the ports—not just in the Catalan countries but around the world—remained remarkably strong, and especially in Barcelona port workers tended to be highly solidaristic with other struggles. Additionally, the international crews of cargo ships were often from the very countries that were most vulnerable to a forced dependence on capitalism, and they had increasingly developed a solidaristic consciousness throughout the travesties of the pandemic and the failing global economy, as many crews were left to starve, abandoned on the high seas with years of unpaid back wages. It also helped that a large proportion of fuel reserves and processing facilities were located at or near ports, in easy reach of port workers.

As for the factories, the Catalan automotive industry had already demonstrated its usefulness at a time when governments and capitalism were leaving people to die: at multiple locations during the pandemic, workers refused to shut down and instead repurposed their factories to manufacture respirators.[2] International, direct relationships of solidarity which proliferated among grassroots movements but were illegible to the experts and technocrats allowed for lines of communication between Catalan factory workers and cooperatives, urban assemblies, and peasant groups in the Global South—from Senegal to Peru—to put out requests for specific machine parts and implements, like oil presses, tractor parts, and medical centrifuges, and to arrange for their shipment.

It should go without saying that another vital move in the global redistribution of the means of survival was the immediate, total, and contemptuous abolition of that most odious of capitalist artifacts, intellectual property. All medicines, all scientific research, all machine blueprints, all code, were immediately made universally accessible, through illegal and forceful means where institutions attempted to maintain exclusive access. Pharmaceutical factories that would not produce freely for everyone were expropriated, university faculties that would not share their research were set on fire. Previously unemployed art students made satirical busts of the putatively great inventors and researchers who had exploited research assistants, taken credit for traditional knowledge, or enclosed useful technology to grow rich while people were dying. The names of many who had expected to be remembered as great men in the annals of history were dragged through the dirt, and will enter popular mythology as contemptibly greedy figures, like the Grinch or King Midas.

Territories that had achieved the autonomy to participate in the initiatives described above often reached self-sufficiency within a decade. The more territories that went down this path, the more the remaining networks of capitalist exploitation desiccated and dried up, the more remaining states lacked the resources and the support to obstruct or annihilate these initiatives, and thus the more other territories could reclaim their autonomy and defend one another against the attempts of capitalist and state institutions to maintain control. The process accelerated exponentially. From the perspective of those in charge, it was an apocalyptic collapse. For the rest of us, though not without extreme hardships, it was the greatest party we'd experienced. Our lives were our own, for the first time.

As more and more territories approached self-sufficiency, cargo shipments declined, coinciding with a decline in available fuel. The unwritten consensus within this growing network of empowered territories was to cap all oil and gas wells in a particular territory as quickly as technically possible, except in cases where doing so would lead to starvation; in such cases, the entire network was committed to helping the territory in question adapt its food infrastructure to shed dependence on fossil fuels.

The obsolescence of container ships, however, has not meant that the seas have been vacated of a human presence. Quite the contrary, the seas and oceans are constantly crisscrossed with human craft, though below the waves a healthy silence reigns, with no more engine noises wreaking havoc on cetacean populations; sea craft are wind-powered. Wherever coastal territories liberated themselves, people expropriated the large sailboats and yachts of the rich and quickly put them to emancipatory purposes. Before the revolution really got under way, the Catalan countries were already plugged into a nascent international anarchist sailors' union.[3] This and similar formations related to the migrant solidarity and the reproductive freedom movements drastically increased their activity, subverting the borders, moving climate and war refugees to safer ground, and supplementing the international shipments of vital supplies.

Now, in this best possible future, after a few decades of revolutionary transformation, global shipping volumes are at only a couple of percent of what they had been at the senescence of global capitalism. People still exchange goods globally, out of pleasure rather than dependence, and since territories are self-sufficient, there is no logic of scarcity and thus no need to regulate exchange within a monetary framework.

Anyone setting sail from a Catalan port, heading to Lápmi or southern Patagonia, is sure to fill up their hold with oranges or olive oil. Such goods are abundant in the fields around Tarragona or Dénia, and anyone who arrives with such treats in a temperate or sub-Arctic latitude is sure to be a popular guest. This practice of distribution, though, does not resemble a quantitative exchange. Those unfortunate peoples who forgot are relearning that hospitality is something that should be offered to anyone who comes in peace; that generosity only creates more generosity.

The interior of the Catalan countries have also experienced a radical transformation. Before the revolution really got under way, the population density was already near the limit of what might be sustainable, and a large part of the surface area was already covered in what technocrats considered "forest." Turning these tree plantations into actual forests is a slow process, though an important one for decreasing atmospheric carbon. In some ecosystems, this means local communities help bring back local oak and beech forests that are fire resistant, promote more local humidity, and over time store a great deal of carbon in the forest floor as well as in the canopy. They also provide food sources to several animal species, enriching the food web in which local human communities take part (acorn flour is also a great substitute, in any year with bad cereal harvests). In ecosystems where pine forests are a better fit, this means shifting back to native pine species and making rigorous use of regenerative pastoralism, with goats and sheep keeping the understory clean and reducing the catastrophic forest fires that are one of the major dangers in our region.

Wetlands sequester much more carbon than forest, and the old wetlands have been restored along all the major river valleys in the lowlands. The reality of catastrophic flooding made it easier to convince people to change houses in places where constructions had to be demolished to make way for recovery. One of the biggest projects was the near-total destruction of the Barcelona airport and the restoration of the wetlands it had been built upon. With the end of mass tourism and jet travel, only one runway was left for small planes. In the Balearic Islands, local residents were ecstatic to reclaim their homes from the disrespectful hordes of tourists, and shutting down the airport to mass traffic was one of the first actions in the transformation. The piers where cruise ships docked were completely sabotaged to prevent any more from arriving and disgorging their unwanted passengers.

During the most delicate years of the transformation, the Catalan countries were lucky to have a large amount of agricultural land and industries that could keep people fed while being repurposed, with enough surplus to send to more arid territories on the other side of the Mediterranean.

One of the most dramatic changes was the abolition of industrial meat production. The factory farming of pigs had dominated much of the agricultural landscape, from the huge warehouses and slaughterhouses where the animals were consigned to miserable lives and deaths to the sprawling monocrop fields dependent on pig slurry for fertilization, such that both plant- and meat-based food products were integrated into the pig industry, as well as dependent on large injections of fossil fuels.

Due to both the dramatic change in resource availability and a spreading ecological consciousness, industrial meat production was shut down with all haste. This also required a radical transformation of farmland. Any animal husbandry had to be integrated with either agriculture or forestry, such that farm animals did not place a demand on the production of grains, but largely fed themselves as they fertilized fields and maintained forests, living in herds and flocks under the open sky. Huge monocrop fields also had to become obsolete, as a reliance on industrial harvesters, chemicals, and petroleum was neither desirable nor practical. Rural space experienced a significant repopulation, such that small gardens near houses have come to produce a large proportion of people's food. Large fields have been broken up and complexified, mixed with fruit and nut trees. Remaining cereal production is done on a smaller scale, harvested with tractors running on biofuels, or in a growing number of projects initiated by highly motivated young people, through the recovery of older ecological technologies and hand tools. Meanwhile, artisanal rice cultivation continues in the wetlands of Tarragona and València.

The result is that the current agricultural system is self-fertilizing, creates a far more diverse and healthy diet, provides good habitat for many bird, reptile, amphibian, insect, and mammalian species, does not pollute, and in fact is carbon negative, in stark contrast to the earlier system.

The shift, on average, of thirty to fifty percent of the urban population to small towns and the countryside is in fact more complex than a simple transfer of people. In fact, affixing a percentage to static categories like urban or rural resident feels like an artifact of the sociological past, given how many people have one foot in the city and one foot in the countryside. Before state power collapsed, there was already a proliferation of anarchist and ecological initiatives recovering the practices and infrastructures necessary for an artisanal, primarily rural, and largely gift-based or solidarity economy. Thanks to these early waves, the models, tools, experiences, and territorial awareness already existed when people faced the need to carry out a rapid transformation of the existing agricultural system. With the abolition of police forces, and once the remaining fascist paramilitaries were defeated, there were no obstacles to occupying the abundant vacant housing units—a byproduct of runaway real estate speculation—fixing up old, abandoned *masies* or farmhouses, or expropriating the *masies* and vacation houses of the wealthy. Easy access to rural housing removed one of the greatest barriers and meant that artisanal and agricultural projects modeled on the anarchist experience could multiply a hundredfold in a matter of years.

These projects involve intense work—but of a sort that is also joyous and healthy, in stark contrast to the farm work of the capitalist past—by small nuclei of permanent rural residents, often organized in queer families that abolish the patriarchal norms. However, again following radical traditions, these nuclei rely on broader networks of friends who come out from the cities at specific times of year—pruning, planting, harvests—both to help out and to enjoy the lavish feasts that always accompany such *jornades* or workdays. Thus, a *masia* that might have eight to ten permanent residents can count on ten to thirty friends who also consider themselves members of the project to come up from the city and spend a few days sharing in the work and bounty. This tradition turns the most grueling moments of agricultural labor into occasions for celebration, as rural residents get to reunite with their semi-urban friends and those friends get to enjoy the countryside and a form of labor that, in small doses and as a contrast to their normal activity in urban workshops, feels wholesome and rejuvenating.

Curiously, a similar transformation took place among the old guard of the obsolete industrial agricultural economy. Losing access to fossil fuels and industrial fertilizers and pesticides in a relatively short time frame, they had neither the ability nor the knowhow to cultivate their large tracts of land. Some of them voluntarily adopted ecological methods—both old and new—that had long percolated at the margins of the agricultural sector, whether preserved by traditionalists or introduced by starry-eyed city folk. But all of them were impressed by the necessity to keep farming to avoid social collapse, and all of them had to seek help in order to cultivate their lands without the noxious aid of petroleum and synthetics. So, similar to the "neo-rurals" and anarchists, they called on their relations. In this case, it was cousins, nieces, nephews, children, and grandchildren who had gone to the city for university and never come back. These larger family networks provided the labor power necessary during key moments like harvests, and the returning youth also tended to erode the conservatism of the old guard of capitalist agriculture, from their patriarchal and xenophobic attitudes to their reticence to ecological practices.

Such rhythms of residence mean that there is far less need to transport food, as more people live in food abundant regions and most food is produced in a small radius. The opposition between city and countryside, a source or an accomplice of many oppressive and state-forming dynamics throughout history, has largely been subverted, as the greatest number of people belong to both spaces.

One struggle that lasted for years and left deep marks in rural regions was the push for racial justice. In the cities, lower class solidarity during the revolt and the abolition of mass media have done a great deal to heal the wounds of racism and marginalize active racists. As for the members of the former upper classes, they are all racists, whether of the progressive or reactionary variety, but they do not matter anymore as they have been fully expropriated and discredited, and no one wants to hear what they have to say.

But in the countryside, industrial farmers inhabited a peculiar borderland between capitalists and workers. They owned means of production and personally enacted the brutal exploitation of migrant workers, mostly racialized people, yet they also worked the fields, were effectively owned by the banks, and tended to live on small margins. In most regions there was little social support for expropriating family farmers.

There was a major push by racialized people and those struggling alongside them to break out of the marginal position the broader society had long consigned them to. That meant access to good housing, which could generally be solved by expropriating the homes of the ultra-wealthy, vacation homes, or vacant units from the real estate bubble. It also meant access to land and food. In some cases, pressure from semi-urban family members succeeded in getting farmers who had exploited migrant labor to give up a part of their lands. In other cases, racialized people expropriated the lands of large agro-companies with the support of "neo-rurals" who were already well established in a territory. Especially in the case of farmers who refused to address their exploitation of migrant workers under capitalism, there remain unhealed wounds that may last for another generation or two. Fortunately, the dominant culture in the cities and in a large part of the countryside is that racism is a value belonging to the old class of plutocrats who nearly got everyone killed. Fighting together in the streets against police violence, evictions, and precarity during the last years of the capitalist regime was key in developing feelings of antiracist solidarity among the youth.

That solidarity was key in the major migrations of the first decade of the transformation. Many immigrants who came to the Catalan countries in the last years of capitalism in search of jobs or fleeing wars chose to go home as communal solidarity economies flourished and wars ended. But many of those who came fleeing the consequences of desertification and ecological collapse, primarily from Africa, chose to stay, and climate refugees are still arriving. Though atmospheric carbon is now decreasing and techniques for mitigating desertification have spread through the vulnerable areas, temperatures still remain more than 2°C over pre-industrial and it is predicted that the climate will remain unstable for another century or two before settling back into the equilibrium of the late Holocene.

Given that the Catalan countries are also experiencing desertification, the social movements had to make a big push to nip any anti-immigration discourses in the bud. Fortunately, in the final years of capitalism solidarity flourished, as tends to happen in disasters. An extensive solidarity infrastructure was developed to help climate refugees get to territories that could support a larger population influx. This included technical aid in setting up agricultural communes in the Castilian interior, which had largely been vacated by capitalist agriculture, or more combative forms of support to accompany people who wanted to move to places like Germany and England, where rainfall was abundant but a larger part of the population doubled down on racist positions during the transformation.

In sum, the abolition of borders and states has allowed migrants and refugees to achieve a dignified survival in the Catalan countries or to use these local territories as a stepping stone to get farther north. There is no central state stifling minority languages and peoples, as Spain had stifled the Catalans, Basques, and others. Without clear borders, and with a proliferation of familial and cultural links to neighboring territories, people come together in the case of conflicts, because society is simply no longer structured in a way that facilitates segmentation or incipient nationalism.

Cities have also transformed significantly. One of the first changes was the multiplication of barricades as people protected themselves from failing state power. The barricades were connected to two subsequent transformations. The first was the proliferation of block assemblies that would become one of the most important units in the subsequent organizational structure, both cellular and federative, that allowed urban residents to reclaim space and shape the city according to their own needs, an impossibility under capitalism. The block assemblies were modeled on well-known earlier experiences of neighborhood assemblies, but at a smaller, intense scale at which everyone could have meaningful participation. The neighborhood assemblies reappeared, but as federative, coordinating spaces for the block assemblies, and themselves federated into a citywide assembly. Nonetheless, block assemblies retained autonomy and in many senses had more power than the citywide assembly, guaranteeing that citywide infrastructure projects had to improve the quality of life for all neighborhoods, rather than establishing sacrifice zones as under the earlier regime. In this way, by locating power at the local level, the pyramid schemes of statist organizing were avoided.

The second rapid transformation linked to the barricades was to make the city impassable to automobile transportation. Not only did the barricades themselves block roads, but the cars themselves were found to be more suitable construction materials for durable blockades, and the gasoline that fueled them was one of the most important resources for popular self-defense against a police force that was increasingly willing to kill anybody to preserve the privileges of their bosses.

Both blocking and requisitioning cars, as it negatively affected the economically privileged classes and protected the lower classes, constituted an important symbolic shift, broadcasting that the city no longer belonged to the wealthy. It was also linked to a rapid decrease in the air and noise pollution that made urban living so hellish. Suddenly, the city became a joyous place of encounter belonging to its own inhabitants. With the abolition of police, property titles were burnt and everyone got access to adequate housing overnight. Even before the noxious hordes of tourists had fled, there had been more than enough vacant units. With property laws gone, artists covered the walls in murals, blunting the ugliness of bourgeois architecture.

Major avenues were opened up to buses, organized workers kept the metro going, and neighbors made sure that all the small streets were fully passable for pedestrians, bikes, and wheelchairs. The next major concern was food. Before the supermarkets had been emptied out, volunteers from each block made contact with workers in the relevant industries to keep production and distribution going in the absence of a monetary economy, also accelerating the aforementioned tradition of mobilizing urban residents to help in harvests and planting. Most streets had half of their paving removed, uncovering the soil beneath. This was subjected to a season or two of fungal and vegetable bioremediation—cleaning up toxins from the capitalist city—and then planted with fruit and nut trees. The few vacant lots that had not already been transformed into urban gardens quickly were. Many neighbors started keeping chickens that pecked and scratched among the fruit trees, also keeping down insect and rat populations. Blocks organized compost collection in coordination with urban gardens and suburban farms, quickly paralleling the system that Kropotkin documented in the nineteenth century that provided Paris with most of its vegetables.[4]

Though many socially useless buildings were demolished early in the transformation—police stations, courthouses, jails and prisons, City Hall—the demolition of excess and substandard housing was delayed until migration flows stabilized and it became clear how many climate refugees would be coming from the Sahel. Once it was known how much extra housing there was, neighborhoods prioritized the demolition of buildings on sites where it would be easiest to rehabilitate the soil and plant gardens. Within two decades, the cities were much more spacious, clean, and beautiful, and more than half of their food came from urban gardens and suburban farms. The rest came from agricultural projects directly supported by semi-urban residents.

Another curious development to affect the cities was the appearance of youth neighborhoods. Surprisingly, the median age of many urban neighborhoods is 20 to 30. This is a result of most people 30 and older moving to small towns and rural projects and their children, sometime in their teenage years, moving to the cities for the educational, social, and cultural opportunities presented. To avoid conflicts with elderly neighbors or neighbors with small children, this youth migration tends to concentrate in a few neighborhoods that had previously been dominated by tourists or wealthy residents, or in abandoned neighborhoods of substandard housing. Some such neighborhoods have become known for non-stop partying, others have become studious university cities, and still others have been fixed up and turned into habitable artistic creations far more imaginative than anything produced by the professional artists of the capitalist regime.

Universities are supported insofar as they are useful to the communities that produce the resources that keep them in operation. That means free education for any who seek it, and research that reflects the needs of communities. Researchers who worked for big industries or developing war and policing technologies—from anthropologists to physicists—have been defrocked and ostracized from academic communities. A great deal of research focuses on climate adaptation and horizontal technologies.

One major scientific intervention is the algae farms used to purify water in several major cities and also used to produce a fast-growing, high-energy biofuel that, combined with carbon capture in the power plants, draws a significant amount of carbon out of the atmosphere. Another is the floating seaweed farms far out in the Mediterranean that help purify the water and also constitute a new food source. One variety of seaweed is added to the diets of sheep and cattle, drastically reducing the amount of methane they emit.[5]

Industry likewise has undergone a vast transformation corresponding to the needs of the communities that support them. Delinked from the global network of production in pursuit of accumulation, factories were either abandoned or taken over by their workers, and in the latter case they were quickly repurposed to meet social needs. The principal forms of production that continue at an industrial scale are for medicines and medical supplies, textiles, and machine parts, though all of these are produced at a much lower volume. As such, most factories are only running a couple days a week, and the workers have also effectuated substantial changes to slow down the pace and increase worker comfort and safety. However, there is no need to ignore that factories are intrinsically designed to put the interests of machinery over the interests of their human attendants. This is no Stakhanovite socialism that prioritizes productivity and pretends that all attendant problems have been solved with a change in management. Factories continue to be nasty places. As such, those who still work in them are either those who feel an affinity for large machines or volunteers who put up with the noise and the artificial environment for a few days every month in order to produce things they want their communities to have access to.

Regarding declining volumes in the three aforementioned categories, the reasons are fairly straightforward. Given that every person has access to a healthy and diverse diet and to preventive medicine, and that air and water pollution, food toxins, and repetitive stress injuries have dropped almost to zero, the major causes of death and disease under the capitalist regime have been largely eliminated. Practically the only reasons for surgical interventions and pharmaceuticals today are preexisting conditions from the capitalist days, residual pollution, genetic disorders, tropical diseases spread by warming, and accidents.

With the decline of synthetics, most textiles are made from flax, cotton, or wool. Textiles now are produced to a higher quality so they last longer, and there has been a marked move away from the middle class ideal of expressing individuality through a large wardrobe. On the other hand, most clothes now are made in artisanal workshops (using textiles from the factories), so fashion is in fact more individualized—often crafted by a friend or acquaintance—rather than mass produced.

As for machine parts, under capitalism they were made to break down and be thrown away, and nowadays they are made to last, and whenever possible repaired. Machines that are frequently produced are those related to the textile and medical industries, those related to agriculture like oil presses or flour mills, hand tools, vehicles used in public transportation like buses and trains, and energy and communications infrastructure. Because industrial production decreased in volume so drastically, workers in different regions coordinate to complement one another and spread production out over a larger network. The point is to find a balance between decreasing the distance components need to be transported, and spreading out the industrial infrastructure so that no one region has to host all the factories that would be needed to produce a train or radio transmitter from start to finish.

As for inorganic raw materials like metals, the vast majority comes from scrapping and recycling. Social movements are pushing for a moratorium on mining, and the truth is that very few communities want mines in their own backyard. Fortunately, capitalism left a great deal of waste, so there is a relative abundance of raw materials already above ground. And a major area of scientific research in the last decade has been the development of synthetics that are neither toxic nor petroleum derivatives.

Popular science, meanwhile, has honed the art of recycling and repurposing. Related to this trend, most production is carried out in small workshops that produce everything from clothes to bicycles to books to building materials. In such workshops, people can develop their creative and technical abilities in an environment that they shape, free to dawdle, experiment, and innovate, as they build and design things that are useful to their communities. Unlike factories, the workshops are also able to carry out vital repair activities, keeping everything from jackets and shoes to motors and stoves in good order.

Communications have undergone a major change, with rocket launches and thus satellites falling by the wayside. The night sky is gradually being reclaimed for stargazers. There is still a strong shared desire for global communication, and such a communication network is vital for adapting to global problems as well as spreading solidarity and overcoming conflict. Radio and telephone have regained a fair deal of currency, and internet is maintained through a combination of preexisting fiber optic networks and newly deployed mesh networks or wireless wide area networks. Internet volume is significantly reduced, both due to the lower supply of electricity and because of the end of capitalist boredom. The biggest drivers of internet usage today are international friendships and the sharing of science and news articles. The new internet infrastructure is fully encrypted and as secure as possible, with no more data mining companies, no more government spies, no more seizures of servers. So far, the recycling of computer and phone hardware has been sufficient, given falling demand, and the production of new replacement parts is pending scientific research into improved recycling techniques or synthetics, given the general moratorium on metals mining.

The energy that powers the communications infrastructure, as well as the factories, workshops, hospitals, and transportation networks, comes from a variety of sources. Around the end of the capitalist regime, the breakdown of electricity production in Catalunya (the most populous of the Catalan countries) was roughly 12 percent from hydroelectric, 6 percent from wind energy, 4 percent from other renewables, 50 percent from nuclear, and 28 percent from combined cycle or cogeneration power plants (mostly using natural gas or biomass as the fuel source).[6] All of these sources underwent major changes.

Engineers at Catalunya's two nuclear power plants were convinced to bring the reactors offline on a schedule that would leave the smallest quantity of highly radioactive waste, meaning that the plants were still producing electricity for several years. Some combined cycle and cogeneration plants transitioned to burn biofuels, while most were shut down in the first year. Solar cells that were already produced before the end of capitalism were quickly redistributed where they were most needed. Industrial wind turbines were dismantled where they were deemed to be particularly harmful to migratory birds or where local farmers demanded it; others were left operational. As electricity needs are stabilized at a new low, biologists, geologists, and local communities are beginning to survey all the hydroelectric dams to determine which ones are most detrimental to river and mountain ecosystems, as a first step towards dismantling them.

Biofuel power plants burn agricultural and residential residues. As electricity usage was brought down, no agricultural land had to be wasted growing fuel for power plants anymore. The plants are fitted with carbon capture systems. The resulting biochar is spread on the fields to improve soil quality or used for high carbon concrete and other building materials.[7]

On balance, the electricity grid in the Catalan countries generates negative carbon emissions thanks to the end of natural gas and petroleum, and the drastic reduction of overall energy usage. Without that reduction, negative emissions would have been impossible, and food shortages would have been caused by extreme demand for biofuels.

Electricity usage was brought down to less than a quarter of the consumption levels at the end of capitalism within a few years. Electricity consumed by industry and the service sector fell immediately with the end of the capitalist economy. Bringing down the electricity needs of the domestic sector required a more thorough transformation as houses were modified to improve insulation and allow for more passive heating and cooling. Small-scale wind turbines and solar panels using tin-based perovskite cells and newer synthetic designs were installed on many rooftops, as well as solar skins on south-facing façades. Where space permits, these systems are plugged into kinetic and gravity batteries, as well as newer, even more ingenious ways of storing energy, thus eliminating the need for destructive chemical batteries and allowing residential blocks to unplug from the grid. This in turn allows the electrical grid to be more concentrated, reducing distribution loss.

While train, metro, and bus networks run off the electricity grid, a small amount of biofuels are produced for farm tractors and trucks in rural regions where the public transportation infrastructure is non-existent. Secondary roads are maintained with gravel or high carbon concrete, but the *autopistes* (interstates or motorways) have been left to degrade. Early on, many highway bridges were sabotaged to slow down the military mobilizations aimed at crushing the movements. Given those associations (militarism, car traffic, the fast life), people have been reticent to bring them back.

A final important question is that of home heating and cooking. While electricity and passive solar are more common in denser urban areas, ever improved stove designs made from recycled metals have proven revolutionary in small towns and rural areas. The stoves allow for cooking and heating, they filter out particle pollution, and they are designed to burn agricultural residues like almond shells and tree prunings, gathered branches and small logs. With tens of millions of olive, almond, apple, cherry, plum, orange, walnut, and other trees throughout the Catalan countries, such residues are abundant.

So what does your territory look like in a happier future, and how does it connect with the one I just described?

One thing I want to make clear is that my proposal is not for a specific vision or final state, but for a method of producing futures. If, for example, you favor responses to the ecological crisis that center more horizontal, situated technologies (that which capitalism does not recognize as technology) or more engineered, interventionist technologies, that's fine. The point is not to build consensus around a blueprint, an impossible proposition on a global scale. Rather, the point is to situate ourselves in a territory, to converse and build relationships with that territory and its other inhabitants, to defend ourselves against those who would annihilate or exploit us, and with all the others grow into something healthier. The vision I describe above is somewhere between my ideal (moderated as a best-case scenario departing from the current hell) and an acknowledgment of where other people are coming from and what they are likely to fight for. In other words, my ideal contains the ideals of other people that are contradictory to my own. As an anarchist, I want to live in a communism inhabited by other people, not by drones toeing the party line.

JUSTICE AND RECONCILIATION: MAKING SURE MASS MURDER DOESN'T PAY

As we have seen throughout this book, the ecological crisis is not a future problem. It is with us right now. Every year, millions of people are being killed and dozens of species are going extinct due to the actions of powerful people and institutions.

A fundamental part of any realistic response to the ecological crisis is to confront these injustices and take on questions of reparations and reconciliation. How do we deal with those responsible? How do we make up for the harm caused? How do we keep these oppressions from being perpetuated? Powerful institutions frequently talk of accountability, but it is clear that destroying the planet and its inhabitants is still a richly rewarded activity. To name just one recent example, the CEO of multinational Rio Tinto was forced to resign in a show of contrition after his company was caught destroying 46,000-year-old aboriginal sacred sites belonging to the Puutu Kunti Kurrama and Pinikura peoples in Australia in the course of iron mining, yet he was simultaneously lavished with a $18.6 million payout, double the previous year.[8]

Concepts like human rights, self-determination, legal process, and redress of grievances are codified into the current global system. Since the end of World War II, there have been numerous official attempts to correct episodes of extreme oppression and brutal atrocities. Nonetheless, the atrocities carried out by the most powerful actors are never punished, and even those injustices that are called to account tend to continue after cosmetic changes. In this era when "climate justice" is a buzzword, that means we need to be every bit as cautious towards institutions that engage with processes of reconciliation as we are towards those that remain silent.

A review of some historically significant processes of reconciliation, reparation, and punishment might at least help us identify useful and counterproductive elements for the road ahead.

The archetypal process can be found in the Nuremberg Trials of 1945–1949, a key moment in the development of international law in the post-war period. Some 200 defendants were charged with war crimes at Nuremberg, with the main trial against the leaders of the Nazi regime resulting in 19 convictions (with 12 death sentences), and three acquittals. For the most part, those convicted were political and military leaders responsible for the execution of civilians and prisoners of war. Plundering—not respecting the property rights of conquered populations—was another thing the Nuremberg judges and prosecutors took quite seriously. The systematic acts of genocide perpetrated by the Nazis entered the trials only tangentially, though genocide as a legal concept was developed in response, with much legal wrangling between the US and the USSR, of course, to keep their preferred forms of genocide out of the definition.[9]

A couple of defendants in the main trial bear examining. Karl Dönitz was the leader of the Germany Navy, and he was convicted of unrestricted naval warfare—meaning the slaughter of tens of thousands of civilians on the open seas—but he was not punished for that offense *because the US military had done the same thing.*[10]

Here we can identify an early limitation of this model of international courts preserving human rights. Like political power, judicial power flows from the barrel of a gun, and any effective court requires the backing of a government, and with it, a military. Today, the International Criminal Court (ICC) in the Hague can only apply its legal standards to countries that have bowed to the military dominance of its backers, primarily NATO. This is why it can prosecute and imprison war criminals from Serbia and Uganda, but it is utterly useless for responding to human rights abuses in Jerusalem or Los Angeles. With the unequal application of punishment, that which is not punished is normalized, so the kinds of abuses systematically carried out by the court's political, military backers become invisible or acceptable. In this sense, the statist, punitive model can make things worse.

This international inequality also parallels the racism of criminal justice within a country. Those who are punished are portrayed as thugs from marginalized territories. Harm, therefore, is not what is carried out by the powerful. The Netherlands, which hosts the ICC, will never have to answer for the many genocidal acts it carried out trying to hold onto its colonies. We are instead taught to look for harm coming from lower down the hierarchy, whether these are racialized inner cities on a domestic scale, countries in the process of Europeanization, as in the Balkans, or countries on the global margins, like Uganda.

Hans Fritzsche was another defendant in the main trial at Nuremberg, and one of the few who was acquitted. He was the head of the news division of the Nazi propaganda service. The Nazis could not have achieved any of what they did without an effective news service, but one of the specific ways that liberal regimes achieve social control is through the alienation of ideas and actions. Media companies are therefore free to organize genocide and oppression on an ideological terrain, and should even be congratulated in the process for promoting free debate. An example of this principle in practice, the *New York Times* played a crucial role in preparing the US invasion of Iraq, responsible for the widespread distribution of information that we knew at the time was false.[11] The fact that not only did the editors suffer no consequences for their direct complicity in over a million deaths, but that most people still take the publication seriously, shows a further harm that can be caused by the principles enshrined at Nuremberg.

Another implicit principle established at Nuremberg is that mass murder is acceptable if it is a result of the pursuit of profit and not the direct objective of whomever carries it out. By and large, the American and British architects of Nuremberg protected German bankers and industrialists from prosecution. There were only three trials to target German industrialists: those against the Flick group, IG Farben, and Krupp. These were largely symbolic trials and what was being prosecuted was primarily these companies' use of slave labor and their plundering of the property of conquered populations, in other words not respecting the property laws that were standard in other industrialized nations. The fact of profiting off a genocidal regime was not put on trial. After all, many North American, British, and French companies also profited under Nazi rule.

In the first two cases, about half the defendants were acquitted, and those who were convicted got short sentences, with the longest at eight years. In the Krupp trial, against the company largely responsible for arming the German military, the longest sentence was 12 years, but after two and a half years they were all released and control of the company was even returned to Alfried Krupp.

The attitude of the new world leaders towards the atrocities of World War II went beyond tolerance. The US recruited and relocated over 1,600 Nazi scientists and technicians who would be militarily useful in the Cold War. For their part, the Soviet Union relocated 2,200 specialists, also looking for advantage. Many of these had participated in unspeakable acts, and all of them had supported the Nazi drive for conquest.

What were the favorable consequences of Nuremberg? The fact of severely punishing at least some of those responsible—the fact that death camp guards had to go on the run and were sometimes imprisoned when caught, even though they were *simply following orders*—helped bring about a change in society. Though our understanding of human rights is contradictory and full of double standards, there is at least an attempt to place human rights before the duty of obedience to authority. The effectiveness of this change can be seen in its omissions. Those forms of complicity with atrocities that were forgiven at Nuremberg are some of the most frequent ways that atrocities are carried out today—encouraging the worldviews that make such atrocities inevitable, profiting off of them, causing harm for the sake of scientific progress, or allowing harm to occur through bureaucratic processes that do not at any stage require someone to pull a trigger or flip a lever. Meanwhile, the more blunt methods of the *Einsatzgruppen*, the death squads, are discouraged outside of *zones of exception*, territories pushed outside the pale of civilization by the War on Terror or colonial geographies of dispossession that condemn certain regions as naturally violent.

As for antisemitism, though it was barely addressed at Nuremberg it became a focus of subsequent prosecutions and judicial processes. Antisemitism is still a major danger today, but support for it has decreased substantially in much of Europe and even right-wing parties usually have to mute such discourses. Punishment of the perpetrators of the Holocaust and the reparation of wealth stolen from Jewish people by the Nazis is a factor in this change. Again, we can measure the change by looking at the omissions. In the mainstream, the very low figure of 6 million victims is frequently given for the Holocaust, omitting, among many others, Roma and Sinti victims. Across the board, these populations have been given scant attention. The result? Throughout Europe, politicians can openly advocate for institutional or vigilante violence against these communities with no consequences, mainstream media systematically talk about them in racist ways, and in the 1990s and 2000s, Germany and France deported thousands of Roma to Romania and Bulgaria, and in Germany as well as Romania they have been subjected to lethal lynchings.[12] Incidentally, mass deportation is one of the offending criteria listed in legal definitions of genocide, and was certainly an important precursor action during the Holocaust.

On balance, a punitive legal response to genocide has had mixed results. Social values have shifted and there is widespread support for the sentiment of "Never Again," but the kinds of atrocities carried out by the authorities who doled out the punishment have become normalized.

Currently, no one is talking about punishing those who are responsible for the ecological crisis and all the death that has gone with it. At most, there are proposals of fining fossil fuel companies. But this is not a punishment, this is simply not letting them profit quite as much from all the misery they cause. So let us entertain the idea of actually holding them responsible.

The execution of Nazi leaders is one of the least controversial acts of justice in history, even among those who oppose the death penalty. Yet suggesting that oil company executives should be executed for knowingly causing tens of millions of deaths, profiting off it, and obstructing social and legal responses, will strike many people as horrific and disproportionate.

By definition, punitive justice is winner's justice.

People today whose survival is currently guaranteed by racial capitalism will go along like good citizens with this chain of atrocities that, in terms of scale, is at least one order of magnitude more lethal than World War II, many of them will agree with the newscasters and the politicians who warn of the twin dangers of fascist and antifascist extremists, and then later, after watching the latest historical documentary, will ask themselves how so many "good Germans" could have let the Holocaust happen.

And yet inevitably, whether ten or a hundred years from now, when most people recognize that today's politicians, millionaires, industry leaders, and many scientists as well were in fact mass murderers, they will play the facile ethical game of asking whether it would be, hypothetically, justified to go back in time to kill some of these monsters, if it might make a difference in affecting the course of history.

This is another aspect of punitive justice: it takes the pursuit of justice out of our hands and turns us into spectators. And when we cannot right the wrongs we see all around us, the typical response is to stop seeing them.

I would argue that these politicians and millionaires have no inviolable worth after everything they have done, and that directing our rage against them is in fact a healthy and just response. However, following an anarchist ethics, I would warn against looking to Nuremberg, not for their sake, but for ours. And this is because any group of people who organize the systematic punishment of their enemies inevitably become monsters, no matter how monstrous those they are punishing are.

Now let us look at justice processes that partially or completely removed the option of punishment from the very beginning. The most famous of these is the Truth and Reconciliation Commission (TRC) of South Africa, in which offenders from any side were promised amnesty if they came forward and confessed to the atrocities they had carried out. Some 20,000 people were identified as victims of human rights abuses, with 3,000 victims identified due to applications for amnesty by the perpetrators.[13] The purpose of this process was to prioritize getting the truth about what happened, and also to move on in a spirit of forgiveness.

The focus of this process was on the final few decades of bloody repression of the anti-apartheid movement. One problem with such a focus is that it leaves out the greater context: that of three centuries of bloody colonization by the Dutch and British and widespread land theft by the Boers. At the end of the process, South Africa still had to beg for

IMF loans while British capitalists still speculated internationally with unimaginable wealth derived in large part from colonialism, white Boers still controlled much of the best lands and owned mines and profitable infrastructure while many Blacks still lived in extreme poverty. And when these impoverished inhabitants rise up in protest, as they did in the Soweto Township to bring the apartheid regime to an end, police are still sent in to shoot them, although now many of those cops are Black.

In other words, there were no real reparations, and the shift in power relations was minimal. The apartheid government got to determine the conditions for its departure. "Success in the constitutional negotiations depended to a large degree on making a deal with the previous regime" and the "minions" of apartheid "still dominated the police, army, and civil service."[14] A new class of Black politicians entered the political elite, and thanks to good old-fashioned corruption some of them got to enter the lower strata of the economic elite, which remains largely white. Perhaps more importantly, South Africa remains plugged into all the structures of global capitalism that arose over the course of European colonization of the world. The representational politics of having Black political leaders serves to obscure these power relations more than to change them.

What about the goal of learning the truth within the narrow focus on repression in the final decades of apartheid? Many families did get to learn specifically how their loved ones died, which is invaluable. And the hearings were designed in a way to honor and center the victims, rather than centering the dispensers of the law, as in a courtroom. I cannot help but think, though, that the narrative of *learning the truth of what happened* is facile, more for the benefit of the so-called international community or for researchers desiring solid documentation. Can anyone doubt that in Soweto, in Sharpeville, people knew exactly the truth of the situation, even if they did not know the specific date, the specific police station, where their neighbors were killed, or exactly who gave the order to open fire on their protest?

Furthermore, the Commission's framing encouraged a pernicious *both sides*-ism, that cynical liberal maneuver of insisting that both sides in a conflict committed regrettable acts. While interpersonal conflicts often do display a dynamic whereby each person's errors aggravate and impel a further round of errors, and therefore the "both sides" portrayal appeals to common sense, it rarely scales up. In a clear case of oppressor and oppressed, framing the extreme acts that the oppressed carry out to win their freedom as comparable to the extreme acts oppressors do to foreclose that freedom is nothing but support for oppression. The British and the Dutch did not stumble over into South Africa innocently and they did not begin plundering the locals because the South Africans had caused them any offense. The TRC did try to break with the victor's justice of Nuremberg. They can be lauded for recognizing that the ends do not justify the means, as ends and means are inseparable. And yet, while movements for liberation do need to be wary of adopting oppressive tactics themselves, that this should happen in a dialogue with their oppressors only erases the reality of the situation they fought against. Furthermore, the principle of "proportionality" established by the TRC considered it more justifiable for apartheid agents to assassinate Blacks involved in armed resistance than those who remained peaceful.

On balance, the South African model is interesting because it helps us imagine a format for justice based on sharing stories and creating paths for healing, but more than that I believe it serves as a cautionary tale for how oppressive institutions can easily recuperate subversive ideas and hollow them out. Ultimately, there can be no healing if the power relations at the root of the oppression are not fixed. And time and again we have seen that representational politics are not capable of changing oppressive power relations, but on the contrary, work in their service. For me, the final word is that most Black survivors of apartheid were not content with the TRC, and many shared the sentiment, "No reconciliation without reparation."[15]

In Guatemala, the Historical Clarification Commission (CEH) of 1995–1997 studied the genocide against Mayan peoples that occurred during the 36-year civil war and resulted in over 200,000 deaths, with many more tortured, raped, and injured. Appointed by the United Nations, the Commission did not have the authority to carry out prosecutions and its findings were barred from having judicial repercussions. Rather, its mandate was to make recommendations and to investigate and document the genocide. Purportedly, the goal was to establish the truth of what happened in order to allow for healing, but this seems like a whitewash, as the Commission was prohibited from naming names or attributing responsibility.

Restitution for victims of the genocide was paid from a minuscule fund set aside by the Guatemalan government, and was characterized by recipients as "symbolic" or "miserable."[16] A small number of military leaders and commandos were eventually prosecuted, though few sentences are actually being applied.

It is, perhaps, significant that the principal architect of the framework of international law represented by the United Nations and thus the CEH, the United States, benefited from and helped to organize the civil war and attendant genocide. The CIA, US Army, and corporations like the United Fruit Company worked with the European-descended Guatemalan elite to launch a coup and carry out a genocidal war against those who resisted, in order to maintain control and continue reaping windfall profits from the extractive economy.[17] The CEH did nothing to change the dispossession of Mayan peoples in Guatemala nor to end the predatory relationship the US government and corporations maintain with Latin America.

Canada also got on the reconciliation bandwagon after a 2006 legal agreement admitted the widespread genocidal practices of residential schools, by which the government kidnapped and abused Indigenous youth and attempted to eradicate their language and culture from the nineteenth century through the 1990s. Simultaneous to waving a flag of peace and healing, the Trudeau government continued aggressively pushing extractivist projects on First Nations land.

In this context, Michif-Cree feminist and anarchist Tawinikay criticizes the concept of reconciliation as "a state-led smoke screen used to advance a more sophisticated policy of assimilation."

> Official Canadian reconciliation centers on accepting the past, apologizing, and moving forward together. It doesn't necessitate physical reparations for the history of colonization. In fact, it discourages that sort of rhetoric as divisive. Counterproductive. Difficult.
>
> There exists a fundamental problem here, because settler-colonialism doesn't exist in the past. Its violence is pervasive and ongoing, right now, tonight, everywhere we look. Reconciliation is the erasure of this current settler-colonial violence.
>
> Reconciliation—as a term—is about resolving a conflict, returning to a state of friendly relations. It can also mean the bringing together of two positions so as to make them compatible.
>
> Decolonization—on the other hand—is about repealing the authority of the colonial state and redistributing land and resources. It also means embracing and legitimizing previously repressed Indigenous worldviews.

> Decolonization isn't a light word. We have to think about what colonization is to understand it: the complete administrative and economic domination of a people and place. Repealing that is a big deal.
>
> Nevertheless, you will often see these two words thrown around almost interchangeably, especially in the university context where folks using them aren't actually actors in struggle. I would argue that this is inappropriate. …
>
> Canada was created in order to govern, exploit, and expand the territories swindled, settled, and stolen from Indigenous peoples of this land. That wasn't a by-product, it was its primary function. It still is. It always will be. It can't escape that.
>
> So how can the Canadian state reconcile with Indigenous peoples? They certainly can't "go back" to a state of friendly relations because there never existed such a time. Reconciliation can only mean eliminating the conflict by enmeshing Indigenous and settler communities, which is the second version of that definition that I shared, making conflicting positions compatible.
>
> This means assimilating Indigenous peoples by having them give up their claim to sovereignty in exchange for the promise of economic equality within Canada. And it means Canadian people get to devour Indigenous ideas and symbols into their own settler stories, their own canadiana. This is the only path possible under the Canadian state. …
>
> Point 7 [of Canada's 10 Points of Official Reconciliation] … says that, consultation is an aspiration, but that the control of land supposedly held by Indigenous peoples can be overridden in any situation beneficial to the state of Canada.
>
> Indigenous peoples, even under the banner of reconciliation, do not have the right to say no to the state of Canada. The right to say no is critical to the realization of sovereignty, of consent, of freedom.[18]

This devastating critique underscores what we have seen in the failings of institutional attempts to achieve justice: there can be no justice if the power relations that generated oppression and unleashed atrocities are not dismantled.

With regard to settler societies, Tawinikay suggests that we could speak of reconciliation between Indigenous and settlers when we abolish capitalism and the state: "the new relationship to place and focus on interdependence will give settlers a chance to genuinely form a new connection to this land themselves."[19]

No justice will be forthcoming if we do not acknowledge that the ecological crisis is victimizing hundreds of millions of people right now, and that the non-human victims of this crisis are also worthy of being remembered, protected, and where possible, avenged.

APOCALYPTIC UTOPIAS NOW: PRESENT STEPS IN THE NETWORKING OF LOCAL AND GLOBAL SPACES

The alternative to ecocide is not a simplistic plan for carbon reduction on a schedule that will not be met, and even if it were could not avert a catastrophe that is already upon us.

The alternative is the networking of thousands of empowered territories, each one approaching the horizon of a complex imagined future that prioritizes the health and happiness of that territory and its inhabitants. To achieve that, we do not have to await the timetables of economists and engineers advising us how to incentivize powerful, murderous institutions to make the needed changes at their own pace, without rocking the boat too much.

What we need to do instead is to find the seeds of those empowered territories all around us, nurture them, cultivate them, and link them in a rhizomatic web that will become stronger than any institution of the ecocidal system that is already falling to ruin, strong enough so that as many of our living communities as possible—human and non-human, all enmeshed—will survive the upheavals and disturbances as our planet tries to heal from centuries of abuse.

Finding those seeds, cultivating them, will look different in every territory. But after talking with friends across the world, I think we can trace an outline for action that will be useful in different contexts by asking ourselves four simple questions. These questions, in conversation with those around us, can be an important step in creating empowered communities.

- *Who is already fighting?*
- *What seedlings do we have?*
- *What risks do we face?*
- *Who can we help?*

When we look around us to see who is already fighting, we are looking for accomplices and also breaking with the mentality of passively awaiting solutions. And by *fighting* I mean struggling for survival, in the broadest sense of the term. Survival is never individual, it is communal, it is physical, emotional, and intergenerational. It includes people trying to stop a pipeline that would poison their water supply, to stop an airport that would bring in floods of tourists and destroy their traditional relation with their territory; people trying to preserve their language or their heirloom crops against pressures to modernize; people fighting against evictions or trash incinerators, trying to create a more livable city; people crossing borders in search of a better life. Fighting for survival does not include people launching careers off their activism. It does not include people developing and patenting technologies that, taken out of context, might seem to make a small part of the problem better but in the current context will just concentrate more wealth into fewer hands.

Taking inventory of our seedlings means identifying sources of strength that have shown to be viable right now, and that with more support might grow exponentially, serving as inspirations or starting points for a new way of living, like the weed that breaks the pavement or the first-generation tree that makes way for a thriving forest. Maybe we take part in a labor organization that might be able to launch infrastructures of international solidarity in a moment of deepening crisis. Maybe we live in a territory that has remained relatively fertile, with a good number of projects that point the way to local food sovereignty; or maybe we live in a desertifying area but take part in projects that help the local ecosystem and human communities to adapt and mitigate the damage. Maybe we run a hacking collective or open technology initiative capable of sharing useful tools and designs on a global scale. Maybe our region has a combative street movement capable of wrecking capitalist infrastructure projects or blocking government policies that threaten us. What links can we establish between projects and what conversations can we have within these projects in order to let more people know they serve as a model for a better world, and to be ready when they need to grow rapidly and pull more weight?

We also need to be aware of what specific risks we face in our territory. Do we live in a food desert? How far away are food and fuel supplies, should the gas stations and supermarkets go empty? What infrastructures do we lack for growing and processing our own food? What repressive strategies and capabilities is the local government likely to use against us, and what kind of organized paramilitary forces exist in our territory? Are we at acute risk of desertification, wildfires, sea level rise, or tropical diseases? Do we have access to the seeds that we will need for next year's planting, or are they all warehoused by agribusiness? How do the people around us react to crisis? When the pandemic hit, did they rush to empty out the stores, or did they think of ways to protect one another?

Finally, who can we help? What living communities—human and non-human—in our territory are most at risk, most in need of additional resources in order to adapt and protect themselves? What infrastructures and resources do we have—especially considering those given to us by global capitalism—that people in other territories lack? How can we build bonds of communication and solidarity with those other territories so we can help them get what they need to survive?

Where we live and who we are will result in very different answers to these questions. No institution on the planet is capable of assimilating all those knowledges in an intelligent way. It is up to us to answer these questions and create our own plans.

It might help to suggest a few lines of inquiry relating to the most basic, broad differences. People from colonized territories have different historical and present examples to look to for deciding how to take charge of their calls for solidarity, the way they build support, and the way they understand reparations. The Zapatistas and the Kurdish liberation movement both put out calls for international solidarity and received important material and moral support. Key elements of their campaigns included inviting supporters to their territories to learn firsthand about their struggles; defining how people from outside could support them and thus eliminating one of the most central and oppressive characteristics of charity; asking those supporters to bring their experiences back to a significantly larger group of people in their home territories and to build support there; and challenging and influencing the ideologies of liberation movements in colonizing countries. Especially in the case of the Zapatistas, there was also an important element of asking people in wealthier countries to raise money and help the local communities achieve the material basis for autonomy in contradiction to the forced dependence of global capitalism.

Other examples include the solidarity campaign that targeted companies doing business with the South African apartheid regime with boycott or sabotage (including an extensive firebombing campaign in Germany and the Netherlands that forced multiple companies to divest), or international solidarity with Palestine focusing on boycotts, divestment, and sanctions. On a smaller scale, but very recently, there was the call put out by First Nations fighting against land grabs in territories occupied by the Canadian state to help blockade transportation infrastructure in the midst of a powerful wave of protests that shut down highways, ports, and rail lines across the country.

To simplify, salient questions include how to communicate the needs of the colonized territory, how to build international relationships strong enough to provide a framework for real solidarity, and how to self-define the struggle and break the habits of paternalism and charity that people from colonizing societies are trained in.

A series of articles by John Severino, all available on theanarchistlibrary.org, detail one model for creating international solidarity based on creating and collectivizing personal relationships between communities in struggle, in this case between anarchists and Indigenous communities within the United States, Bolivia, and Chile.

For our part, people from colonizing countries might want to consider recent critiques of allyship, such as "Accomplices Not Allies" by Indigenous Action Media or "Critique of Ally Politics."[20] Breaking with paternalistic charity models is fundamental; if we are not learning from the people we are supporting, it isn't solidarity. But there are other pitfalls as well.

Tawinikay speaks to one of them:

> Saying you support Indigenous sovereignty doesn't mean backing every Indigenous person on every project. There are plenty of Indigenous misogynists and ladder-climbing politicians out there, and you don't do me any favours by helping them gain power. Fight for liberatory ideas, not for nations or bloodlines.
>
> We do this all the time. There are Indigenous people out there who oppose pipelines and those who support them, but we align ourselves with the resistance, so we are making choices already. Own it. It's okay. It's good to fight for the land and for freedom.
>
> … Understand what struggles are about and know who is participating in them. Get to know those people. Build relationships. Build meaningful relationships outside of the occupation, as friends. It can't start from a place of white guilt. Don't get swept up in your own settler redemption story.[21]

As for the imperative of rooting ourselves in a territory, this can be extremely difficult if we have no such roots to start with, if we come from a culture that has been completely modernized, and especially if we are settlers in a settler state. Unless they are well thought out in historical perspective and with great sensitivity to all the other living communities around us, attempts to "reconnect with our roots" will only end up reproducing alienation.

Knowing the Land is Resistance (https://knowingtheland.com) provides an interesting model for an initiative by settlers living "between Lake Ontario and the Niagara Escarpment" to learn from and connect with local woodlands; to spread ecological skills and awareness with plant walks, seed harvesting, zines, and workshops; to develop an "anarchist ecology" that critiques "the ecology of management from a distance, and of remote expertise, that sees itself as fundamentally separate from the land, inhabiting a present without a past or future" and instead offers "starting points for an anti-colonial, anti-authoritarian way of connecting with the land [that is] rooted in relationships, deep listening, urban ecology, re-enchanting, and unexpertness"; and to build active solidarity with ongoing Indigenous struggles to defend and recover their land and their ways of life.

Because of a history we did not create but must nonetheless situate ourselves within, settlers need to make clear that it is more important for us to defend the land than to own it. This means sabotaging destructive infrastructure and helping block ecocidal megaprojects without the expectation that doing so will win us ownership or some other claim to the land.

Plenty of past projects that arose from European socialism and focused on reclaiming a connection with the land became complicit with colonialism when carried out in a settler context, including Zionist kibbutzim such as Degania or Sointula, a commune set up by Finnish socialists on the territory of the 'Namgis First Nation, usurped within the confines of British Columbia.[22]

As for specific considerations for Indigenous and racialized people moving forward, of course I'm not qualified to offer any, but I hope that the revolutionaries, communities, organizations, and initiatives that have also spoken in these pages can provide inspiration relevant to many different contexts.

One thing that I think is true wherever we are is that we can get to know the earth beneath our feet, the history that towers behind us, the overlapping communities that surround us. Wherever we live, there we can find life, and with life, struggle. Let us connect on that basis, and fight for a world fit for living things, not for the ossified institutions begging us to save them, to rejuvenate them again with one more round of life-sapping reforms.

Capitalism is dying. It has been dying for a very long time, and still it keeps growing. Colonialism has already changed its masks. The state has stood on the precipice of collapse before. Why keep pretending? Why bite our tongues? Let us put them out of their misery, before it is too late for all of us. We don't need them. There is another world right here. It has been with us the whole time.

Notes

All internet references were last consulted May 8, 2021 unless otherwise noted. All block texts are from personal communications with the author, unless otherwise noted. Interviews originally in Spanish and Catalan were translated by the author, interviews originally in Portuguese were translated by Zenite.

1. A WIDE-ANGLE VIEW

1. Jeff Berardelli, "How Climate Change Is Making Hurricanes More Dangerous," *Yale Climate Connections*, July 8, 2019, https://yaleclimateconnections.org/2019/07/how-climate-change-is-making-hurricanes-more-dangerous.
2. Alejandra Borunda, "The Science Connecting Wildfires to Climate Change," *National Geographic*, September 17, 2020, www.nationalgeographic.com/science/2020/09/climate-change-increases-risk-fires-western-us.
3. Orlando Milesi and Marianela Jarroud, "Soil Degradation Threatens Nutrition in Latin America," *Inter Press Service*, June 15, 2016, https://reliefweb.int/report/world/soil-degradation-threatens-nutrition-latin-america; United Nations, "The United Nations Convention to Combat Desertification: A New Response to an Age-Old Problem," 1997.
4. Dim Coumou, Alexander Robinson, and Stefan Rahmstorf, "Global Increase in Record-Breaking Monthly-Mean Temperatures," *Climatic Change* 118: 771–782 (2013), https://doi.org/10.1007/s10584-012-0668-1.
5. WWF, "Living Planet Report 2020," September 10, 2020.
6. Virginia Institute of Marine Science, "Study Shows Continued Spread Of 'Dead Zones': Lack Of Oxygen Now A Key Stressor On Marine Ecosystems," *ScienceDaily*, August 15, 2008, www.sciencedaily.com/releases/2008/08/080814154325.htm.
7. D. Pimentel and K. Hart, "Pesticide Use: Ethical, Environmental, and Public Health Implications," in A. W. Galston and E. G. Shurr (eds.), *New Dimensions in Bioethics* (Boston, MA: Springer, 2001).
8. Andrew S. Lewis, "Tax on Superfund Polluters Gets New Life," *NJ Spotlight News*, April 22, 2021, www.njspotlight.com/2021/04/superfund-sites-rep-frank-pallone-nj-reintroduces-tries-again-tax-polluters-site-cleanup-he-first-introduced-2005.
9. Erena Calvo, "La minería deja una huella tóxica en la salud de la Sierra Minera de Cartagena," *El Diario*, August 26, 2017, www.eldiario.es/murcia/industrializacion-pesadilla-sierra-minera-cartagena_1_3223100.html.

10. C. Milesi, C. D. Elvidge, J. B. Dietz, B. T. Tuttle, R. R. Nemani, and S. W. Running, "A Strategy for Mapping and Modeling the Ecological Effects of US Lawns," www.isprs.org/proceedings/XXXVI/8-W27/milesi.pdf.
11. Lester R. Brown, "Paving the Planet: Cars and Crops Competing for Land," Earth Policy Institute, February 14, 2001.
12. Oliver Milman, "'Invisible Killer': Fossil Fuels Caused 8.7m Deaths Globally in 2018, Research Finds," *The Guardian*, February 9, 2021, www.theguardian.com/environment/2021/feb/09/fossil-fuels-pollution-deaths-research.
13. World Health Organization, "Climate Change and Health," February 1, 2018, www.who.int/news-room/fact-sheets/detail/climate-change-and-health.
14. World Health Organization, "Diarrhoeal Disease," May 2, 2017, www.who.int/news-room/fact-sheets/detail/diarrhoeal-disease.
15. Jos Lelieveld, Klaus Klingmüller, Andrea Pozzer, Ulrich Pöschl, Mohammed Fnais, Andreas Daiber, and Thomas Münzel, "Cardiovascular Disease Burden from Ambient Air Pollution in Europe Reassessed Using Novel Hazard Ratio Functions," *European Heart Journal* 40(20): 1590–1596 (May 2019), https://doi.org/10.1093/eurheartj/ehz135.
16. Philip J. Landrigan, Richard Fuller, et al., "The Lancet Commission on Pollution and Health," *The Lancet* 391(10119) (October 2017).
17. Francesca Mataloni, Chiara Badaloni, et al., "Morbidity and Mortality of People Who Live Close to Municipal Waste Landfills: A Multisite Cohort Study," *International Journal of Epidemiology* 45(3): 806–815 (June 2016).
18. Internal Displacement Monitoring Centre, "Internal Displacement from January to June 2019," September 12, 2019, www.internal-displacement.org/sites/default/files/inline-files/2019-mid-year-figures_for%20website%20upload.pdf.
19. International Organization for Migration, "Migration and Climate Change," no. 31 (2008).
20. WWF, "Half of Plant and Animal Species at Risk from Climate Change in World's Most Important Natural Places," March 14, 2018, www.worldwildlife.org/press-releases/half-of-plant-and-animal-species-at-risk-from-climate-change-in-world-s-most-important-natural-places.
21. Holly Jean Buck, *After Geoengineering: Climate Tragedy, Repair, and Restoration* (London: Verso, 2019), p. 217.
22. Dim Coumou and Alexander Robinson, "Historic and Future Increases in the Global Land Area Affected by Monthly Heat Extremes," *Environmental Research Letters* 8(3) (2013).
23. Jim Dobson, "Shocking New Maps Show How Sea Level Rise Will Destroy Coastal Cities by 2050," *Forbes*, October 30, 2019, www.forbes.com/sites/jimdobson/2019/10/30/shocking-new-maps-show-how-sea-level-rise-will-destroy-coastal-cities-by-2050.
24. Kurt Campbell, Jay Gulledge, et al., *The Age of Consequences: The Foreign Policy and National Security Implications of Global Climate Change* (Washington, DC: Center for A New American Security, 2007), p. 9.

25. Ryan Prior, "The Water Is So Hot in Alaska it's Killing Large Numbers of Salmon," August 17, 2019. https://edition.cnn.com/2019/08/16/us/alaska-salmon-hot-water-trnd/index.html.
26. Timothy M. Lenton, Johan Rockström, et al., "Climate Tipping Points: Too Risky to Bet Against," *Nature* 575: 592–595 (2019).
27. Helmholtz Association of German Research Centres, "The Forests of the Amazon Are an Important Carbon Sink," November 11, 2019, https://phys.org/news/2019-11-forests-amazon-important-carbon.html.
28. Thomas E. Lovejoy and Carlos Nobre, "Amazon Tipping Point," *Science Advances* 4(2) (2018).
29. Lenton et al., "Climate Tipping Points: Too Risky to Bet Against."
30. WWF, "Half of Plant and Animal Species at Risk from Climate Change in World's Most Important Natural Places," www.wwf.org.uk/updates/half-plant-and-animal-species-risk-climate-change-worlds-most-important-natural-places-0.
31. Yangyang Xu and Veerabhadran Ramanathan, "Well Below 2°C: Mitigation Strategies for Avoiding Dangerous to Catastrophic Climate Changes," *Proceedings of the National Academy of Sciences* 114(39): 10315–10323 (2017). Cited in David Spratt and Ian Dunlop, "Existential Climate-Related Security Risk: A Scenario Approach" (Melbourne: Breakthrough National Centre for Climate Restoration, 2019).
32. *Ibid.*, p. 6.
33. University of Bristol, "Ever-Increasing CO_2 Levels Could Take Us Back to the Tropical Climate of the Paleogene Period," July 30, 2018, www.bristol.ac.uk/news/2018/july/co2-levels-paleogene-period.html.
34. Megan Trimble, "Air Pollution Causes 8.8 Million Extra Deaths a Year," *US News & World Report*, March 12, 2019, www.usnews.com/news/national-news/articles/2019-03-12/air-pollution-causes-88-million-extra-deaths-worldwide-each-year-study-says; and Landrigan et al., "The Lancet Commission on Pollution and Health."
35. Scott Crow, *Black Flags and Windmills: Hope, Anarchy, and the Common Ground Collective* (Oakland, CA: PM Press, 2011).
36. Neil Smith, "There's No Such Thing as a Natural Disaster," https://libcom.org/library/there's-no-such-thing-natural-disaster-neil-smith.
37. Frans de Waal, *Are We Smart Enough to Know How Smart Animals Are?* (New York: W. W. Norton Company, 2016).
38. Brooke Jarvis, "The Insect Apocalypse is Here," *New York Times Magazine*, November 27, 2018, www.nytimes.com/2018/11/27/magazine/insect-apocalypse.html.
39. Pamela Boycoff, "By Learning to Think Like a Puffin, This Conservationist Has Saved Seabirds around the World," *CNN*, October 14, 2020, https://edition.cnn.com/2020/09/30/world/puffins-maine-kress-c2e-scn-spc-intl/index.html.
40. Isabelle Groc, "Furry Engineers: Sea Otters in California's Estuaries Surprise Scientists," *The Guardian*, August 14, 2020, www.theguardian.com/

environment/2020/aug/14/natures-furry-engineers-sea-otters-in-california-estuaries-surprise-scientists-aoe.

41. Robin McKie, "How Sea Otters Help Save the Planet," *The Guardian*, July 10, 2016, www.theguardian.com/environment/2016/jul/10/sea-otters-global-warming-trophic-cascades-food-chain-kelp.
42. Robin Smith, "The Future of a Fog Oasis," *Scientific American*, August 19, 2011, https://blogs.scientificamerican.com/guest-blog/the-future-of-a-fog-oasis.
43. *Ibid.*
44. *Ibid.*
45. Leah Asmelash, "Australia's Indigenous People Have a Solution for the Country's Bushfires. And it's Been around for 50,000 Years," *CNN*, January 12, 2020, https://edition.cnn.com/2020/01/12/world/aboriginal-australia-fire-trnd/index.html.
46. Mike Gouldhawke, April 28, 2021, https://twitter.com/M_Gouldhawke/status/1387494937214607365?s=19.
47. Jim Robbins, "Orcas of the Pacific Northwest Starving and Disappearing," *The New York Times*, July 9, 2018, www.nytimes.com/2018/07/09/science/orcas-whales-endangered.html.
48. David D. Smits, "The Frontier Army and the Destruction of the Buffalo: 1865–1883," *Western Historical Quarterly* 25(3): 312–338 (Autumn 1994), p. 316.
49. Benjamin I. Cook, Ron L. Miller, and Richard Seager, "Amplification of the North American 'Dust Bowl' Drought through Human-Induced Land Degradation," *Proceedings of the National Academy of Sciences of the United States of America* 106(13): 4997–5001 (March 2009).
50. Peter Gelderloos, "Commoning and Scarcity," *Tides of Flame*, 2012, https://theanarchistlibrary.org/library/peter-gelderloos-commoning-and-scarcity.
51. E. P. Thompson, *Whigs and Hunters: The Origin of the Black Act* (New York: Pantheon Books, 1975); Peter Linebaugh, *Red Round Globe Hot Burning* (Oakland, CA: University of California Press, 2019).
52. James C. Scott, *The Art of Not Being Governed* (New Haven, CT: Yale University Press, 2009), pp. 25–26.
53. Negreverd, *12 Historias Ludditas* (Zaragoza: Moai, 2015).
54. André Léo, *La Guerre Social*, ed. Michèle Perrot (n.l.: Le Passager clandestin, 2010 [1871]).
55. Alfredo M. Bonanno, *Armed Joy*, trans. Jean Weir (London: Elephant Editions, 1988 [1977]); Anonymous, "At Daggers Drawn," trans. Jean Weir (London: Elephant Editions, 1998 [no date found for original]).
56. Frank Kitson, *Low Intensity Operations: Subversion, Insurgency, and Peacekeeping* (London: 1971); James Hughes, "Frank Kitson in Northern Ireland and the 'British Way' of Counterinsurgency," *History Ireland* 22(1) (2014).
57. Alexander Dunlap, "'Agro sí, mina no!' The Tía Maria Copper Mine, State Terrorism, and Social War by Every Means in the Tambo Valley, Peru," *Political Geography* 71: 10–25 (2019); Alexander Dunlap, "Wind, Coal, and

Copper: The Politics of Land Grabbing, Counterinsurgency, and the Social Engineering of Extraction," *Globalizations* 17(4): 661–682 (2020).

58. Giovanni Arrighi, *The Long Twentieth Century: Money, Power and the Origins of our Times* (London: Verso, 2010 [1994]).
59. Mikhail Bakunin, *Statism and Anarchy*, trans. Marshall S. Shatz (Cambridge: Cambridge University Press, 1990 [1873]).
60. Fredy Perlman, *Against His-story, Against Leviathan* (Detroit, MI: Black & Red, 1983).
61. Josep Gardenyes, *23 Tesis en torno a la revuelta* (Barcelona: Distri Josep Gardenyes, 2011), p. 19, my translation.
62. Jonathan Watts and Denis Campbell, "Half of Child Psychiatrists Surveyed Say Patients Have Environmental Anxiety," *The Guardian*, November 20, 2020, www.theguardian.com/society/2020/nov/20/half-of-child-psychiatrists-surveyed-say-patients-have-environment-anxiety.
63. Edith Bracho-Sanchez, "Number of Children Going to ER with Suicidal Thoughts, Attempts Doubles, Study Finds," *CNN*, April 8, 2019, https://edition.cnn.com/2019/04/08/health/child-teen-suicide-er-study/index.html.
64. Karien Stronks, Aydın Şekercan, et al., "Higher Prevalence of Depressed Mood in Immigrants' Offspring Reflects Their Social Conditions in the Host Country: The HELIUS Study," *PLoS One*, June 4, 2020; Chesmal Siriwardhana and Robert Stewart, "Forced Migration and Mental Health: Prolonged Internal Displacement, Return Migration and Resilience," *International Health* 5(1): 19–23 (2013).
65. Rob Jordan, "Stanford Researchers Find Mental Health Prescription: Nature," *Stanford News*, June 30, 2015, https://news.stanford.edu/2015/06/30/hiking-mental-health-063015.
66. Sing C. Chew, *World Ecological Degradation: Accumulation, Urbanization, and Deforestation 3000 BC–AD 2000* (Walnut Creek, CA: Altamira Press, 2001), pp. 110–111.
67. Conrad Totman, "Forest Products Trade in Pre-Industrial Japan," in John Dargavel (ed.), *Changing Pacific Forests* (Durham, NC: Forest History Society, 1992), pp. 19–20, cited in Chew, *World Ecological Degradation*, p. 115.
68. Chew, *World Ecological Degradation*, pp. 66–70.
69. J. Donald Hughes, "Ancient Deforestation Revisited," *Journal of the History of Biology* 44: 43–57 (2011).
70. Peter Gelderloos, *Worshiping Power* (Oakland, CA: AK Press, 2017).
71. Scott, *The Art of Not Being Governed.*
72. Richard Grove, *Green Imperialism: Colonial Expansion, Tropical Island Edens and the Origins of Environmentalism, 1600–1860* (Cambridge: Cambridge University Press, 1995).
73. Thompson, *Whigs and Hunters.*
74. Jared Diamond, *Collapse: How Societies Choose to Fail or Succeed* (London: Penguin Books, 2005).

75. Gerry Marten, "How Japan Saved Its Forests: The Birth of Silviculture and Community Forest Management," from "Environmental Tipping Points: A New Paradigm for Restoring Ecological Security," *Journal of Policy Studies (Japan)* 20: 75–87 (July 2005).
76. Ishi Hiroyuki, "Protecting the Forests," March 19, 2018, www.nippon.com/en/features/c03912.
77. Douglas W. Owsley et al., "Biological Effects of European Contact on Easter Island," in C. S. Larson, G.R. Milner (eds.), *In the Wake of Contact: Biological Responses to Conquest* (New York: Wiley-Liss, 1993).
78. Gelderloos, *Worshiping Power.*
79. Woods Hole Oceanographic Institute, "Climate Change Likely Caused Migration, Demise of Ancient Indus Valley Civilization," November 13, 2018, www.whoi.edu/press-room/news-release/climate-change-likely-caused-migration-demise-of-ancient-indus-valley-civilization.
80. Richard Kemeny, "Double Climate Disaster May Have Ended Ancient Harappan Civilisation," *New Scientist*, November 20, 2020, www.newscientist.com/article/2261018-double-climate-disaster-may-have-ended-ancient-harappan-civilisation.
81. Max Engel and Helmut Brückner, "Holocene Climate Variability of Mesopotamia and its Impact on the History of Civilisation," in Eckart Ehlers and Katajun Amirpur (eds.), *Middle East and North Africa: Climate, Culture, and Conflicts* (Leiden: Brill, 2021), pp. 77–113.
82. Chew, *World Ecological Degradation*, 37–38.
83. Gelderloos, *Worshiping Power.*
84. James J. Aimers, "What Maya Collapse? Terminal Classic Variation in the Maya Lowlands," *Journal of Archaeological Research* 15: 329–377 (2007), p. 347.
85. Katrina Armstrong, "The Classic Maya Collapse: The Importance of Ecological Prosperity," *Earth Common Journal* 4(1) (2014); Richardson B. Gill, *The Great Maya Droughts: Water, Life, and Death* (Albuquerque, NM: University of New Mexico Press, 2000).
86. Aimers, "What Maya Collapse?" p. 349.
87. Lázaro Hilario Tuz Chi, "Así es nuestro pensamiento. Cosmovisión e identidad en los rituales agrícolas de los mayas peninsulares," doctoral thesis, Universidad de Salamanca, 2009, pp. 50–51, 54–55.
88. John Severino, "'The Other Gods Were Crying': Stories of Rebellion in the Bolivian Highlands," December 11, 2010, https://theanarchistlibrary.org/library/john-severino-the-other-gods-were-crying.html.
89. Gelderloos, *Worshiping Power.*
90. Stephanie Buck, "The First American Settlers Cut Down Millions of Trees to Deliberately Engineer Climate Change," August 22, 2017, https://timeline.com/american-settlers-climate-change-5b7b68bd9064.
91. "The Firewood Shortage that Helped Give Birth to America," November 15, 2017, www.history.com/news/the-firewood-shortage-that-helped-give-birth-to-america.

92. Patrick Manning, "The Slave Trade: The Formal Demographics of a Global System," in Joseph E. Inikori and Stanley L. Engerman (eds.), *The Atlantic Slave Trade: Effects on Economies, Societies and Peoples in Africa, the Americas, and Europe* (Durham, NC: Duke University Press, 1992).
93. Walter Rodney, *How Europe Underdeveloped Africa* (Baltimore, MD: Black Classic Press, 2011 [1972]); Vandana Shiva, *The Violence of the Green Revolution: Ecological Degradation and Political Conflict in Punjab* (New Delhi: Natraj Publishers, 1989).
94. Fiona Harvey, "World's Richest 1% Cause Double CO_2 Emissions of Poorest 50%, Says Oxfam," *The Guardian*, September 21, 2020, www.theguardian.com/environment/2020/sep/21/worlds-richest-1-cause-double-co2-emissions-of-poorest-50-says-oxfam.
95. Kathryn Yusoff, *A Billion Black Anthropocenes* (Minneapolis, MN: University of Minnesota Press, 2019).

2. FOXES BUILDING HENHOUSES

1. Graham Purchase, *Green Flame: Kropotkin and the Birth of Ecology* (Fordsburg, South Africa: Zabalaza Books, 2013).
2. Reuters, "Timeline: How the World Found Out about Global Warming," November 22, 2012, www.reuters.com/article/us-climate-talks-history/timeline-how-the-world-found-out-about-global-warming-idUSBRE8ALoLE20121122.
3. Oliver Milman, "Oil Firms Knew Decades Ago Fossil Fuels Posed Grave Health Risks, Files Reveal," *The Guardian*, March 18, 2021, www.theguardian.com/environment/2021/mar/18/oil-industry-fossil-fuels-air-pollution-documents.
4. John Cook, Dana Nuccitelli, et al., "Quantifying the Consensus on Anthropogenic Global Warming in the Scientific Literature," *Environmental Research Letters* 8(2) (2013).
5. Michael E. Mann, *The New Climate War: The Fight to Take Back Our Planet* (New York: PublicAffairs Books, 2021).
6. Niall McCarthy, "Oil and Gas Giants Spend Millions Lobbying to Block Climate Change Policies," *Forbes*, March 25, 2019, www.forbes.com/sites/niallmccarthy/2019/03/25/oil-and-gas-giants-spend-millions-lobbying-to-block-climate-change-policies-infographic.
7. Robert McChesney, *Corporate Media and the Threat to Democracy* (New York: 7 Stories Press, 1997); Robert McChesney, *Rich Media, Poor Democracy: Communication Politics in Dubious Times* (New York: The New Press, 1999); Noam Chomsky, *Necessary Illusions: Thought Control in Democratic Societies* (London: Pluto Press,1989).
8. Media and Climate Change Observatory, "Monthly Summaries," issue 9, September 2017, https://sciencepolicy.colorado.edu/icecaps/research/media_coverage/summaries/issue9.html.
9. Lisa Hymas, "Climate Change Is the Story You Missed in 2017. And the Media Is to Blame," *The Guardian*, December 7, 2017, www.theguardian.

com/commentisfree/2017/dec/07/climate-change-media-coverage-media-matters; Kevin Kalhoefer, "Study: ABC and NBC Drop the Ball on Covering the Impact of Climate Change on Hurricanes," September 8, 2017, www.mediamatters.org/tucker-carlson/study-abc-and-nbc-drop-ball-covering-impact-climate-change-hurricanes.

10. Genevieve LeBaron, "Green NGOs Cannot Take Big Business Cash and Save Planet," *The Guardian*, October 1, 2013, www.theguardian.com/environment/2013/oct/01/green-ngos-big-business-naomi-klein.
11. INCITE! Women of Color Against Violence, *The Revolution Will Not Be Funded: Beyond the Non-Profit Industrial Complex* (Durham, NC: Duke University Press, 2017).
12. "The Highest Paid Nonprofit CEOs in 2019," www.causeiq.com/insights/highest-paid-nonprofit-ceos.
13. Donna Laframboise, "The Enormous CEO Salaries Behind Earth Hour," March 28, 2012, https://nofrakkingconsensus.com/2012/03/28/the-enormous-ceo-salaries-behind-earth-hour.
14. United Nations Environment Programme, "The Emissions Gap Report 2012," 2012, p. 2.
15. David G. Victor, Keigo Akimoto et al., "Prove Paris Was More than Paper Promises," *Nature* 548(7665): 25–27 (August 2017).
16. Jonathan Watts, "Domino-Effect of Climate Events Could Push Earth into a 'Hothouse' State," *The Guardian*, August 7, 2018, www.theguardian.com/environment/2018/aug/06/domino-effect-of-climate-events-could-push-earth-into-a-hothouse-state.
17. K. Levin and R. Bradley, "Working Paper: Comparability of Annex I Emission Reduction Pledges," World Resources Institute, February 2010, p. 16.
18. Lenton et al., "Climate Tipping Points: Too Risky to Bet Against."
19. Damian Carrington, "Big Banks' Trillion-Dollar Finance for Fossil Fuels 'Shocking', Says Report," *The Guardian*, March 24, 2021, www.theguardian.com/environment/2021/mar/24/big-banks-trillion-dollar-finance-for-fossil-fuels-shocking-says-report.
20. Buck, *After Geoengineering*, p. 207.
21. Kate Aronoff, "Shell Oil Executive Boasts That His Company Influenced the Paris Agreement," *The Intercept*, December 8, 2018, https://theintercept.com/2018/12/08/shell-oil-executive-boasts-that-his-company-influenced-the-paris-agreement.
22. Laura Terzani, personal communication with the author, July 12, 2021.
23. Alexander Dunlap, "Renewing Destruction: Wind Energy Development in Oaxaca, Mexico," doctoral thesis, Vrije Universiteit Amsterdam, 2017, p. 130. On "property based approaches to counterinsurgency," Dunlap cites Joe Bryan and Denis Wood *Weaponizing Maps: Indigenous Peoples and Counterinsurgency in the Americas* (New York: The Guildford Press, 2015), p. 149.
24. Graham Readfearn, "Australia's Bushfires Have Emitted 250m Tonnes of CO_2, almost Half of Country's Annual Emissions," *The Guardian*,

December 13, 2019, www.theguardian.com/environment/2019/dec/13/australias-bushfires-have-emitted-250m-tonnes-of-co2-almost-half-of-countrys-annual-emissions; Adam Morton and Graham Readfearn, "The Disaster Movie Playing in Australia's Wild Places—and Solutions that Could Help Hit Pause," *The Guardian*, March 5, 2021, www.theguardian.com/environment/2021/mar/05/the-disaster-movie-playing-in-australias-wild-places-and-solutions-that-could-help-hit-pause.

25. Maddie Stone, "Solar Panels Are Starting to Die, Leaving Behind Toxic Trash," August 22, 2020, www.wired.com/story/solar-panels-are-starting-to-die-leaving-behind-toxic-trash.
26. Andrea Brock, Benjamin K. Sovacool, and Andrew Hook, "Volatile Photovoltaics: Green Industrialization, Sacrifice Zones, and the Political Ecology of Solar Energy in Germany," *Annals of the American Association of Geographers*, 2021.
27. Alice Kim, "Puna Geothermal Works to Plug Final Well as Lava Approaches," Hawai'i Groundwater & Geothermal Resources Center, May 21, 2018, www.higp.hawaii.edu/hggrc/puna-geothermal-works-to-plug-final-well-as-lava-approaches.
28. "Radioactive Leaks Found at 75% of US Nuke Sites," June 21, 2011, www.cbsnews.com/news/radioactive-leaks-found-at-75-of-us-nuke-sites.
29. WISE Uranium Project, "Depleted Uranium Inventories," April 21, 2008, www.wise-uranium.org/eddat.html.
30. Dunlap, "Renewing Destruction," p. 77.
31. *Ibid.*, p. 80.
32. *Ibid.*, pp. 88–90.
33. *Ibid.*, p. 85.
34. *Ibid.*, pp. 95–97.
35. *Ibid.*, pp. 99–100, 102.
36. *Ibid.*, pp. 90-91.
37. *Ibid.*, pp. 81–82.
38. *Ibid.*, p. 83.
39. *Ibid.*, pp. 109–111.
40. *Ibid.*, pp. 107–108.
41. *Ibid.*, pp. 131–132.
42. *Ibid.*, p. 43.
43. BNEF, "German Coal Mine to Be Reborn as Giant Pumped Hydropower Battery," March 17, 2017, https://about.bnef.com/blog/german-coal-mine-to-be-reborn-as-giant-pumped-hydropower-battery.
44. Andrea Diaz and Steve Almasy, "Canadian Government Says Taxes from Reapproved Pipeline Expansion Project Will Go towards Green Energy," June 19, 2019, https://edition.cnn.com/2019/06/18/americas/canada-trans-mountain-pipeline-expansion/index.html.
45. Jonathan Watts, "Deep-Sea 'Gold-Rush': Secretive Plans to Carve Up Seabed Decried," *The Guardian*, December 9, 2020, www.theguardian.com/environment/2020/dec/09/secretive-gold-rush-for-deep-sea-mining-dominated-by-handful-of-firms.

46. Sandra van Niekerk, "Resource Rich and Access Poor: Securing a Just Transition to Renewables in South Africa," in Edouard Morena, Dunja Krause, and Dimitris Stevis (eds.), *Just Transitions: Social Justice in the Shift Towards a Low-Carbon World* (London: Pluto Press, 2020), p. 196.
47. Greenpeace, "Oil in the Cloud," May 19, 2020, www.greenpeace.org/usa/reports/oil-in-the-cloud.
48. Max Opray, "Nickel Mining: The Hidden Environmental Cost of Electric Cars," *The Guardian*, August 24, 2017, www.theguardian.com/sustainable-business/2017/aug/24/nickel-mining-hidden-environmental-cost-electric-cars-batteries.
49. Annie Kelly, "Children as Young as Seven Mining Cobalt Used in Smartphones, Says Amnesty," *The Guardian*, January 19, 2016, www.theguardian.com/global-development/2016/jan/19/children-as-young-as-seven-mining-cobalt-for-use-in-smartphones-says-amnesty.
50. Matthew Rozsa, "Elon Musk Becomes Twitter Laughingstock after Bolivian Socialist Movement Returns to Power," October 20, 2020, www.salon.com/2020/10/20/elon-musk-becomes-twitter-laughingstock-after-bolivian-socialist-movement-returns-to-power.
51. Telesur, "Evo Morales: Lithium was the Reason for the Coup in Bolivia," November 10, 2020, www.telesurenglish.net/news/Morales-Lithium-Was-the-Reason-for-the-Coup-D-Etat-in-Bolivia-20201110-0014.html.
52. Oliver Balch, "The Curse of 'White Oil': Electric Vehicles' Dirty Secret," *The Guardian*, December 8, 2020, www.theguardian.com/news/2020/dec/08/the-curse-of-white-oil-electric-vehicles-dirty-secret-lithium.
53. Nina Avramova, "Climate Change Is Already Here, and Heat Waves Are Having the Biggest Effect, Report Says," November 29, 2018, https://edition.cnn.com/2018/11/28/health/global-climate-change-and-health-report-intl/index.html.
54. Bronson W. Griscom, Justin Adams, et al., "Natural Climate Solutions," *PNAS* 114(44): 11645–11650 (2017).
55. John Vidal, "'Large-Scale Human Rights Human Rights Violations' Taint Congo National Park Project," *The Guardian*, November 26, 2020, www.theguardian.com/world/2020/nov/26/you-have-stolen-our-forest-rights-of-baka-people-in-the-congo-ignored.
56. Stephen Corry, "The Two Faces of Conservation," August 2015, www.survivalinternational.org/articles/3396-the-two-faces-of-conservation.
57. *Ibid.*
58. Azhar Qadri and Hannah Ellis-Petersen, "'They are Custodians of the Jungle': Anger as Kashmiri Nomads' Homes Destroyed," *The Guardian*, December 4, 2020, www.theguardian.com/world/2020/dec/04/they-are-custodians-of-the-jungle-anger-as-kashmiri-nomads-homes-destroyed.
59. Dunlap, "Renewing Destruction," p. 114.
60. "Open Letter to the Lead Authors of 'Protecting 30% of the Planet for Nature: Costs, Benefits and Implications," https://openlettertowaldronetal.wordpress.com.

61. Laura Terzani, "The Black Mambas Anti-poaching Unit – Are There Any Elephants in the Room? A Decolonial-Feminist Political Ecology Analysis of Conservation in the Greater Kruger National Park," MA dissertation, University of Sussex, 2019, p. 7.
62. Buck, *After Geoengineering*, p. 203.
63. Jerry Redfern, "'No One Explained': Fracking Brings Pollution, Not Wealth, to Navajo Land," *The Guardian*, April 4, 2021, www.theguardian.com/us-news/2021/apr/04/navajo-nation-fracking.
64. Jonathan Watts, "China Plans Rapid Expansion of 'Weather Modification' Efforts," *The Guardian*, December 3, 2020, www.theguardian.com/world/2020/dec/03/china-vows-to-boost-weather-modification-capabilities.
65. Nathaniel Rich, "Climate Change and the Savage Human Future," *The New York Times*, November 16, 2018, www.nytimes.com/interactive/2018/11/16/magazine/tech-design-nature.html.
66. Scott, *The Art of Not Being Governed.*
67. Perlman, *Against His-story, Against Leviathan.*
68. Chew, *World Ecological Degradation*, p. 96.
69. Colin Allen and Michael Trestman, "Animal Consciousness," in Edward N. Zalta (ed.), *The Stanford Encyclopedia of Philosophy* (Winter 2020 edition), https://plato.stanford.edu/entries/consciousness-animal/#hist, §3.
70. Silvia Federici, *Caliban and the Witch: Women, the Body and Primitive Accumulation* (New York: Autonomedia, 2004), 158–159.
71. Simon Worral, "There Is Such a Thing as Plant Intelligence," *National Geographic*, February 21, 2016, www.nationalgeographic.com/science/article/160221-plant-science-botany-evolution-mabey-ngbooktalk.
72. Thích Nhất Hạnh, *Peace Is Every Step: The Path of Mindfulness in Everyday Life* (New York: Bantam, 1992), p. 100.
73. Sulak Sivaraksa, *Seeds of Peace: A Buddhist Vision for Renewing Society* (Berkeley, CA: Parallax Press, 1992), p. 7.
74. Gabriel Kuhn, *Liberating Sápmi: Indigenous Resistance in Europe's Far North* (Oakland, CA: PM Press, 2020), p. 55.
75. Niillas Somby in Kuhn, *Liberating Sápmi*, pp. 57, 66.
76. Quoted in Kuhn, *Liberating Sápmi*, p. 7.
77. Jacqueline Hookimaw-Witt, "Keenebonanoh keemoshominook kaeshe peemishikhik odaskiwakh [We Stand on the Graves of Our Ancestors]: Native Interpretations of Treaty #9 with Attawapiskat Elders," graduate thesis for Trent University, Peterborough, Ontario, 1997, quoted in Mike Gouldhawke, "Land as a Social Relationship," *Briarpatch*, September 10, 2020, https://briarpatchmagazine.com/articles/view/land-as-a-social-relationship.
78. Gouldhawke, "Land as a Social Relationship."
79. Cecilia Nowell, "In the Navajo Nation, Anarchism Has Indigenous Roots," *The Nation*, September 25, 2020, www.thenation.com/article/activism/anarchism-navajo-aid.

80. Megan Mayhew Bergman, "'We're at a Crossroads': Who Do the Fish of Hawaii Belong to?" *The Guardian*, August 26, 2020, www.theguardian.com/environment/2020/aug/26/hawaii-fish-waters-native-commercial-fishers.
81. Vine Deloria Jr, *Custer Died for Your Sins: An Indian Manifesto* (Norman, OK: University of Oklahoma Press, 1988 [1969]), pp. 102,104.
82. Giacomo D'Alisa, Federico Demaria, and Giorgos Kallis (eds.), *Degrowth: A Vocabulary for a New Era* (Abingdon-on-Thames: Routledge, 2015); Jason Hickel and Giorgos Kallis, "Is Green Growth Possible?" *New Political Economy* 54(4): 469–486 (2020).
83. Federico Demaria, Giorgos Kallis, and Karen Bakker, "Geographies of Degrowth: Nowtopias, Resurgences and the Decolonization of Imaginaries and Places," *Nature and Space* 2(3): 431–450 (2019), p. 432.
84. Frankie Chappell, "Terra Nullius?" April 7, 2020, https://royalsociety.org/blog/2020/04/terra-nullius; Roy Macleod, "Passages in Imperial Science: From Empire to Commonwealth," *Journal of World History* 4(1) (1993).
85. Yusoff, *A Billion Black Anthropocenes*; Linda Tuhiwai-Smith, *Decolonizing Methodologies: Research and Indigenous Peoples* (London: Zed Books, 2002).
86. Wendy Makoons Geniusz, *Our Knowledge Is Not Primitive: Decolonizing Botanical Anishinaabe Teachings* (Syracuse, NY: Syracuse University Press, 2009).
87. Bergman, "'We're at a Crossroads': Who Do the Fish of Hawaii Belong to?"
88. Matthias Monroy, "Social Movements Against the Global Security Architecture!" October 7, 2008, https://digit.site36.net/2008/10/07/social-movements-against-the-global-security-architecture.
89. General Klaus Naumann, General John Shalikashvili, Field Marshal The Lord Inge, et al., "Towards a Grand Strategy for an Uncertain World: Renewing Transatlantic Partnership," Noaber Foundation, 2007.
90. CrimethInc., "Designed to Kill: Border Policy and How to Change It," zine updated into book, CrimethInc., *No Wall They Can Build*, 2017.
91. "Second Indigenous Activist Killed in Honduras This Week," *Al Jazeera*, December 12, 2020, www.aljazeera.com/news/2020/12/31/honduras-sees-second-indigenous-activist-killed-in-in-past-week.
92. Nathalie Butt, Mary Menton, "More than 1,700 Activists Have Been Killed this Century Defending the Environment," *The Conversation*, August 5, 2019, https://theconversation.com/more-than-1-700-activists-have-been-killed-this-century-defending-the-environment-120352 Citing Global Witness report.
93. Jack Guy, "Record Number of Environmental Activists Killed in 2019," July 29, 2020, https://edition.cnn.com/2020/07/29/world/global-witness-2019-defenders-report-scli-intl/index.html.
94. Juan Smith, "Colombia: Ex-Paramilitary Implicates Two U.S. Companies in Murder of Trade Unionists," December 14, 2009, https://nacla.org/news/colombia-ex-paramilitary-implicates-two-us-companies-murder-trade-unionists.

95. Adriaan Alsema, "Colombia Charges 13 Former Chiquita Executives over Hundreds of Murders," *Colombia Reports*, September 1, 2018, https://colombiareports.com/terror-for-profit-colombia-charges-14-former-chiquita-executives.
96. Steven Cohen, "The Supreme Court Needs to Decide: Can Victims Sue Chiquita for Sponsoring Terrorism," *New Republic*, April 10, 2015, https://newrepublic.com/article/121506/supreme-court-should-take-chiquitas-terrorism-sponsorship-case.
97. Colombia Reports, "Civilian Casualties of Colombia's Armed Conflict," July 20, 2019, https://colombiareports.com/civilians-killed-armed-conflict. For a standard example of dishonest reporting, see "Colombia's Guerrilla War Killed 260,000, Report Says," *CBC*, August 2, 2018, www.cbc.ca/news/world/colombia-guerrilla-farc-death-toll-1.4771858.
98. Joe Parkin Daniels, "Colombian Army Killed Thousands More Civilians than Reported, Study Claims," *The Guardian*, May 8, 2018, www.theguardian.com/world/2018/may/08/colombia-false-positives-scandal-casualties-higher-thought-study.
99. "Ijaw Tribe," December 19, 2005, https://onlinenigeria.com/finance/?blurb=669.
100. A. A. Kadafa, "Environmental Impacts of Oil Exploration and Exploitation in the Niger Delta of Nigeria," *Global Journal of Science Frontier Research* 12(3) (2012).
101. BBC, "Nigerian Attack Closes Oilfield," June 20, 2008, http://news.bbc.co.uk/1/hi/world/africa/7463288.stm.
102. Caleb Maupin, "The Niger Delta Avengers Declare War on Western Oil Giants," June 8, 2016, www.mintpressnews.com/niger-deltans-plenty-avenge-yet-us-media-ignores-context-terrorism-oil-companies/217030.
103. Naomi Larsson, "Killed for Defending the Planet: Murder of Environmental Activists Reaches Record High," *Independent*, August 5, 2020, www.independent.co.uk/news/world/europe/environmental-activists-deaths-murder-climate-change-colombia-romania-a9651826.html.
104. Adam Federman, "Revealed: US Listed Climate Activist Group as 'Extremist' alongside Mass Killers," *The Guardian*, January 13, 2020, www.theguardian.com/environment/2020/jan/13/us-listed-climate-activist-group-extremists.
105. Andrea Brock, "Enforcing Ecological Catastrophe at All Costs," *New Internationalist*, November 19, 2020, https://newint.org/features/2020/11/19/enforcing-ecological-catastrophe-all-costs.
106. Andrea Brock, "The Battle of Hambacher Forest," *Red Pepper*, September 20, 2018, www.redpepper.org.uk/the-battle-of-hambacher-forest.
107. Dylan Rodríguez, "Reformism Isn't Liberation, It's Counterinsurgency," *Black Agenda Report*, October 21, 2020, https://blackagendareport.com/reformism-isnt-liberation-its-counterinsurgency; Yannick Giovanni Marshall, "Black Liberal, Your Time Is Up," *Al Jazeera*, June 1, 2020, www.aljazeera.com/opinions/2020/6/1/black-liberal-your-time-is-up.

108. Connor Woodman, "The Infiltrator and the Movement," *Jacobin*, April 23, 2018, https://jacobinmag.com/2018/04/uk-infiltration-secret-police-mi5-special-branch-undercover.
109. Paul Lewis and Rob Evans, "Secrets and Lies: Untangling the UK 'Spy Cops' Scandal," *The Guardian*, October 28, 2020, www.theguardian.com/uk-news/2020/oct/28/secrets-and-lies-untangling-the-uk-spy-cops-scandal.
110. Colze a Colze, "Presentació," April 10, 2017, https://colzeacolzeblog.wordpress.com/2017/04/10/presentacio.
111. Will Potter, *Green Is the New Red: An Insider's Account of a Social Movement Under Siege* (San Francisco, CA: City Light Publishers, 2011).
112. Green Is the New Red, "SHAC 7," www.greenisthenewred.com/blog/tag/shac-7.
113. David Harrison, "Minister Set up Deal to Save Animal Lab," *The Telegraph*, January 21, 2001, www.telegraph.co.uk/news/uknews/1318759/Minister-set-up-deal-to-save-animal-lab.html.
114. Gene Johnson, "Two Face Federal Terror Charge over Train Track Interference," *AP News*, December 1, 2020, https://apnews.com/article/bellingham-washington-british-columbia-seattle-7afd76ecb1ee9bccac7d0bcb7f9ae1b7.
115. Sam Levin, "Revealed: How Monsanto's 'Intelligence Center' Targeted Journalists and Activists," *The Guardian*, August 8, 2019, www.theguardian.com/business/2019/aug/07/monsanto-fusion-center-journalists-roundup-neil-young.
116. Catherine Nolin, Grahame Russell, "Public Letter," November 11, 2020, https://mailchi.mp/rightsaction/public-letter-publication-upcoming-book-blocked?fbclid=IwARoehaY2JNu05DBHgtoesuljK8FCFGOOT-tKxnTxnaHgopI1pXUK9Y9r-gs t.
117. Tracy Glynn, "Did the RCMP Attend a University Book Launch to Stop a Crime?" *NB Media Co-op*, January 22, 2021, https://nbmediacoop.org/2021/01/22/did-the-rcmp-attend-a-university-book-launch-to-stop-a-crime.
118. Oliver Milman, "Facebook Suspends Environmental Groups Despite Vow to Fight Misinformation," *The Guardian*, September 22, 2020, www.theguardian.com/environment/2020/sep/22/facebook-climate-change-environment-groups-suspended; "Facebook Bans Multiple Anarchist and Antifascist Pages," *Indybay*, August 24, 2020, www.indybay.org/newsitems/2020/08/24/18836084.php.
119. Enough 14, "Dutch Police Seized Servers of NoState Tech Collective—North Shore Counter-info, Montreal Counter-info, 325 and Act for Freedom Now," March 31, 2021, https://enoughisenough14.org/2021/03/31/dutch-police-seized-servers-of-nostate-tech-collective-north-shore-counter-info-montreal-counter-info-325-and-act-for-freedom-now-down.
120. Kim Willsher, "Anarchists Hijack Climate March on Day of Violent Protests in Paris," *The Guardian*, September 21, 2019, www.theguardian.com/

world/2019/sep/21/paris-on-high-alert-as-protesters-try-to-revive-gilets-jaunes-movement.

121. Harsha Walia, *Border and Rule: Global Migration, Capitalism, and the Rise of Racist Nationalism* (Chicago, IL: Haymarket Books, 2021).
122. Paul Hockenos, "Germany's Secret Labor Experiment," *The New York Times*, May 9, 2018, www.nytimes.com/2018/05/09/opinion/germans-secret-labor-experiment.html.
123. The Migrant Project, "Migrant Deaths in the Mediterranean Exceed 20,000 since 2014," March 9, 2020, www.themigrantproject.org/mediterranean-deaths-2.
124. Marianna Karakoulaki, "The Invisible Violence of Europe's Refugee Camps," *Al Jazeera*, October 22, 2019, www.aljazeera.com/opinions/2019/10/22/the-invisible-violence-of-europes-refugee-camps; Council of Europe, "Council of Europe's Anti-torture Committee Calls on Greece to Reform its Immigration Detention System and Stop Pushbacks," November 19, 2020, www.coe.int/en/web/cpt/-/council-of-europe-s-anti-torture-committee-calls-on-greece-to-reform-its-immigration-detention-system-and-stop-pushbacks.
125. Anna Giaritelli, "Deadly Crossing: Bodies of 300 Illegal Immigrants Found on US-Mexico Border during Fiscal 2019," *Washington Examiner*, December 11, 2019, www.washingtonexaminer.com/news/deadly-crossing-bodies-of-300-illegal-immigrants-found-on-us-mexico-border-during-fiscal-2019.
126. Aviva Shen, "Why Did it Take a Sterilization Scandal to Retrigger our Outrage over ICE?" *Slate*, September 18, 2020, https://slate.com/news-and-politics/2020/09/ice-sterilization-scandal-outrage-abuse-hysterectomies.html; Maya Manian, "Immigration Detention and Coerced Sterilization: History Tragically Repeats Itself," September 29, 2020, www.aclu.org/news/immigrants-rights/immigration-detention-and-coerced-sterilization-history-tragically-repeats-itself.
127. René Kladzyk, "ICE Detainees Face Systemic Torture at El Paso Area Immigration Facility, New Report Says," *El Paso Matters*, January 5, 2021, https://elpasomatters.org/2021/01/05/ice-detainees-at-el-paso-area-immigration-facility-face-systemic-torture-new-report-says/; Julian Borger, "US Ice Officers 'Used Torture to Make Africans Sign Own Deportation Orders'," *The Guardian*, October 22, 2020, www.theguardian.com/us-news/2020/oct/22/us-ice-officers-allegedly-used-torture-to-make-africans-sign-own-deportation-orders.
128. Thalia Anthony and Stephen Gray ("Was There Slavery in Australia? Yes, it Shouldn't Even Be Up for Debate," *The Conversation*, June 12, 2020, https://theconversation.com/was-there-slavery-in-australia-yes-it-shouldnt-even-be-up-for-debate-140544) describe the process but incorrectly identify the enslavement of Aborigines as ending in the 1950s. For continuity through the 1970s see Sarah Collard, "Class Action Launched against West Australian Government over Indigenous Stolen Wages," *ABC News*, October 18, 2020, www.abc.net.au/news/2020-10-19/wa-government-

faces-class-action-over-stolen-wages/12737046; Norman Hermant, "Seasonal Farm Workers Receiving Less than $10 a Week after Deductions, Investigation Reveals," *ABC News*, February 26, 2016, www.abc.net.au/news/2016-02-25/seasonal-farm-workers-receiving-as-little-as-$9-a-week/7196844.

129. Stephen Charles, "Our Detention Centres Are Concentration Camps and Must Be Closed," *Sydney Morning Herald*, May 4, 2016, www.smh.com.au/comment/our-detention-centres-are-intentionally-cruel-and-must-be-closed-20160504-golr04.html.
130. Deanna Dadusc and Pierpaolo Mudu, "Care Without Control: The Humanitarian Industrial Complex and the Criminalisation of Solidarity," *Geopolitics*, April 17, 2020, pp. 25–26, 30.
131. Robert Mackey, "Trump Boasts about Federal Task Force Killing Anti-Fascist Wanted for Murder in Portland," *The Intercept*, October 15, 2020, https://theintercept.com/2020/10/15/trump-boasts-federal-task-force-killing-antifascist-wanted-murder-portland.
132. Nick Estes, *Our History is the Future* (London: Verso, 2019), p. 54.
133. *Ibid.*, p. 251.
134. Sandy Tolan, "Wounded on the Front Line at Standing Rock, a Protester Refuses to Give Up Her Fight," *Los Angeles Times*, December 22, 2016, www.latimes.com/nation/la-na-standing-rock-wounded-20161222-story.html; Tim Fontaine, "Woman Hurt in Latest Standing Rock Confrontation in 'Serious' Condition," *CBC*, November 22, 2016, www.cbc.ca/news/indigenous/standing-rock-serious-injuries-1.3861705.
135. Estes, *Our History is the Future*, p. 55.
136. Ryan Fatica, "Steve Martínez Jailed Again after Refusing to Testify Before a Federal Grand Jury," *Perilous Chronicle*, March 3, 2021, https://perilouschronicle.com/2021/03/03/steve-martinez-jailed-again-after-refusing-to-testify-before-a-federal-grand-jury.
137. Ward Churchill, *From a Native Son: Selected Essays on Indigenism, 1985–1995* (Cambridge, MA: South End Press, 1996), pp. 256–260.
138. Democracy Now, "Leonard Peltier Speaks from Prison," June 12, 2000, www.democracynow.org/2000/6/12/leonard_peltier_speaks_from_prison.
139. Tamara Starblanket, *Suffer the Little Children: Genocide, Indigenous Nations and the Canadian State* (Atlanta, GA: Clarity Press, 2018).
140. Jaskiran Dhillon and Will Parrish, "Exclusive: Canada Police Prepared to Shoot Indigenous Activists, Documents Show," *The Guardian*, December 20, 2019, www.theguardian.com/world/2019/dec/20/canada-indigenous-land-defenders-police-documents.
141. Brent Patterson, "Secwepemc Land Defender Kanahus Manuel Calls for a National Inquiry into the Ts'Peten/Gustafsen Lake Standoff," *PBI*, December 12, 2020, https://pbicanada.org/2020/12/12/secwepemc-land-defender-kanahus-manuel-calls-for-a-national-inquiry-into-the-tspeten-gustafsen-lake-standoff.
142. Freedom News, "Breaking: Russian Antifascists sentenced to 6-18 Years Prison," February 10, 2020, https://freedomnews.org.uk/breaking-russian-

antifascists-sentenced-to-6-18-years-prison; RURepression, "How the FSB Is Manufacturing a Terrorism Case Against Antifascists in Russia," https://rupression.com/en/how-the-fsb-is-manufacturing-a-terrorism-case-against-antifascists-in-russia.

143. Elizabeth C. Economy, "A Land Grab Epidemic: China's Wonderful World of Wukans," Council on Foreign Relations, February 7, 2012; and Michael Forsythe, "China's Spending on Internal Police Force in 2010 Outstrips Defense Budget," *Bloomberg*, March 6, 2011, www.bloomberg.com/news/2011-03-06/china-s-spending-on-internal-police-force-in-2010-outstrips-defense-budget.html.
144. Agence France-Presse, "China Bomb Attack Kills Four in Suspected Protest Over Development," *The Guardian*, March 23, 2021, www.theguardian.com/world/2021/mar/23/china-bomb-attack-kills-four-in-suspected-protest-over-development.
145. Liu Jianqiang, "How Tiger Leaping Gorge Was Saved," *China Dialogue*, https://chinadialogue.net/en/nature/5923-how-tiger-leaping-gorge-was-saved; Michael Standaert, "With Activists Silenced, China Moves Ahead on Big Dam Project," *The Third Pole*, November 5, 2020, www.thethirdpole.net/2020/11/05/with-activists-silenced-china-moves-ahead-on-big-dam-project.
146. Li You, "Ningxia Conservationists Detained for 'Picking Quarrels,'" *Sixth Tone*, September 11, 2020, www.sixthtone.com/news/1006168/ningxia-conservationists-detained-for-picking-quarrels.
147. It's Going Down, "We Don't Forget: Support Joseph Dibee, Environmentalist Accused of Sabotage," August 20, 2018, https://itsgoingdown.org/we-dont-forget-support-joseph-dibee.
148. Carl Gibson and Steve Horn, "Exposed: Globally Renowned Activist Collaborated with Intelligence Firm Stratfor," December 2, 2013, www.occupy.com/article/exposed-globally-renowned-activist-collaborated-intelligence-firm-stratfor.
149. Ronald Duchin, quoted in Steve Horn, "Here's How the Corporations Defeat Political Movements," July 29, 2013, https://popularresistance.org/heres-how-the-corporations-defeat-political-movements.
150. *Ibid.*
151. *Ibid.*
152. Peter Gelderloos, *The Failure of Nonviolence* (Seattle, WA: Left Bank Books, 2013); Peter Gelderloos, "Riots and Remembrance on the Streets of Barcelona: The Collective Learning of Subversive Techniques," in Alissa Starodub and Andrew Robinson (eds.), *Riots and Militant Occupations: Smashing a System, Building a World—A Critical Introduction* (London: Rowman & Littlefield International, 2018).

3. THE SOLUTIONS ARE ALREADY HERE

1. A. Ananthalakshmi, "Palm Oil to Blame for 39% of Forest Loss in Borneo Since 2000: Study," Reuters, September 19, 2019, www.reuters.com/article/us-palmoil-deforestation-study-idUSKBN1W41HD.

2. Emily Zhao, "Unilever and Nestlé Are Burning Indonesia. Is 'Sustainable Palm Oil' a Con?" November 11, 2019, https://medium.com/the-climate-reporter/unilever-and-nestlé-are-burning-indonesia-is-sustainable-palm-oil-a-con-4a15e3110d1a; "The Copernicus Atmospheric Monitoring Service Tracks Extent and Pollution from Fires across Indonesia," September 20, 2019, https://atmosphere.copernicus.eu/copernicus-atmosphere-monitoring-service-tracks-extent-and-pollution-fires-across-indonesia.
3. Krystof Obidzinski, "Fact File—Indonesia World Leader in Palm Oil Production," *Forest News*, July 8, 2013, https://forestsnews.cifor.org/17798/fact-file-indonesia-world-leader-in-palm-oil-production.
4. Kuhn, *Liberating Sápmi*, p. 40.
5. *Ibid.*, p. 111.
6. Aslak Holmberg in Kuhn, *Liberating Sápmi*, pp. 112, 118.
7. Kuhn, *Liberating Sápmi*, p. 55.
8. Niillas Somby in Kuhn, *Liberating Sápmi*, pp. 58, 66.
9. CrimethInc, "La ZAD: Another End of the World Is Possible," April 9, 2018, https://crimethinc.com/2018/04/09/la-zad-another-end-of-the-world-is-possible-learning-from-50-years-of-struggle-at-notre-dame-des-landes.
10. See https://crimethinc.com/2019/04/23/reflections-on-the-zad-looking-back-a-year-after-the-evictions and https://infokiosques.net/IMG/pdf/SPLASH3-letter.pdf.
11. "Protest Culture: History," 2007, archived at https://web.archive.org/web/20071013213726/www.protestculture.org/history.html.
12. "Reclaim the Streets," *Do or Die* 6: 1–10 (1997), archived at www.doordie.org.uk.
13. CrimethInc. "The Forest Occupation Movement in Germany," March 10, 2021, https://crimethinc.com/2021/03/10/the-forest-occupation-movement-in-germany-tactics-strategy-and-culture-of-resistance.
14. "Poland: Anarchists Successfully Stop Allotment Gardens Eviction," *Freedom News*, November 18, 2020, https://freedomnews.org.uk/poland-anarchists-successfully-stop-allotment-gardens-eviction.
15. Some of these histories are discussed in Gord Hill, *The 500 Years of Resistance Comic Book* (Vancouver: Arsenal Pulp Press, 2010).
16. Wilder Utopia, "Fracking in New Brunswick: Elsipogtog First Nation Takes a Stand," December 7, 2013, www.wilderutopia.com/environment/energy/fracking-new-brunswick-elsipogtog-first-nation-takes-stand.
17. Tawinikay, "Reconciliation Is Dead: A Strategic Proposal," *It's Going Down*, February 15, 2020, https://itsgoingdown.org/reconciliation-is-dead-a-strategic-proposal.
18. See www.tinyhousewarriors.com/#about.
19. Hill, *The 500 Years of Resistance Comic Book*, pp. 74, 87.
20. CrimethInc, "Don't See What Happens, Be What Happens," January 29, 2017, https://crimethinc.com/2017/01/29/dont-see-what-happens-be-what-happens-continuous-updates-from-the-airport-blockades.

21. Kevin Ridder, "The Appalachian Pipeline Resistance Movement: We're Not Going Away," *Appalachian Voice*, October 28, 2020, https://appvoices.org/2020/10/28/the-appalachian-pipeline-resistance-movement.
22. Hugh Fearnley-Whittingstall, "If the UK Government Won't Stop Industrial Fishing from Destroying Our Oceans, Activists Will," *The Guardian*, February 26, 2021, www.theguardian.com/commentisfree/2021/feb/26/if-the-uk-government-stop-industrial-fishing-oceans-activists-greenpeace.
23. Vandana Shiva, *Stolen Harvest: The Hijacking of the Global Food Supply* (Cambridge, MA: South End Press, 2000); Shiva, *The Violence of the Green Revolution.*
24. Jonathan Watts, "1% of Farms Operate 70% of World's Farmland," *The Guardian*, November 24, 2020, www.theguardian.com/environment/2020/nov/24/farmland-inequality-is-rising-around-the-world-finds-report.
25. Jonathan Watts, "Third of Earth's Soil Is Acutely Degraded due to Agriculture," *The Guardian*, September 12, 2017, www.theguardian.com/environment/2017/sep/12/third-of-earths-soil-acutely-degraded-due-to-agriculture-study.
26. World Future Council, "How Does Agriculture Contribute to Climate Change?" October 21, 2012, www.worldfuturecouncil.org/how-does-agriculture-contribute-to-climate-change.
27. PG, "Coordinating a Gift Economy," *Fifth Estate* 395 (Winter 2016).
28. Stephen Heyman, "Soul Fire Farm's Leah Penniman Explains Why Food Sovereignty Is Central in the Fight for Racial Justice," *Vogue*, July 2, 2020, www.vogue.com/article/soul-fire-farm-leah-penniman-why-food-sovereignty-is-central-in-the-fight-for-racial-justice; Leah Penniman, *Farming While Black: Soul Fire Farm's Practical Guide to Liberation to Liberation on the Land* (White River Junction, Vermont: Chelsea Green Publishing Company, 2018).
29. Alex Wilson, "Becoming Intimate with the Land," *Briarpatch*, September 10, 2020, https://briarpatchmagazine.com/articles/view/becoming-intimate-with-the-land.
30. Rafa Arques, *La Fuerza del Fuego* (Alcoi, València: Editorial Milvus, 2019), pp. 15, 35–38, 43–44.
31. *Ibid.*, pp. 52–53, 68, 86–87, 90.
32. Robbie Corey Boulet, "Despite Snags, Ethiopia Scales up Massive Tree-Planting Campaign," June 5, 2020, https://phys.org/news/2020-06-snags-ethiopia-scales-massive-tree-planting.html.
33. Jason Burke, "Young Men Take Up Arms in Northern Ethiopia as Atrocities Fuel Insurgency," *The Guardian*, March 8, 2021, www.theguardian.com/world/2021/mar/08/atrocities-insurgency-ethiopia-tigray.
34. Gelderloos, *Worshiping Power.*
35. Anthony Langat, "The Traditions that Could Save a Nation's Forests," *BBC*, November 4, 2020, www.bbc.com/future/article/20201103-the-indigenous-wisdom-that-can-save-forests-from-destruction.
36. *Ibid.*

37. Lou del Bello, "In India, Indigenous Tribes Clash with the Government Over Trees," *Undark*, January 6, 2020, https://undark.org/2020/01/06/india-indigenous-trees.
38. Amrit Dhillon, "Millions of Forest-Dwelling Indigenous People in India to Be Evicted," *The Guardian*, February 22, 2019, www.theguardian.com/world/2019/feb/22/millions-of-forest-dwelling-indigenous-people-in-india-to-be-evicted.
39. Cathy Watson, "Indian Farmers Fight against Climate Change Using Trees as a Weapon," *The Guardian*, October 29, 2016, www.theguardian.com/global-development-professionals-network/2016/oct/29/indian-farmers-fight-against-climate-change-using-trees-as-a-weapon.
40. Hannah Ellis-Petersen, "Nationwide Farmers' Strike Shuts Down Large Parts of India," *The Guardian*, December 8, 2020, www.theguardian.com/world/2020/dec/08/nationwide-farmers-strike-shuts-down-large-parts-of-india.
41. Pranav Jeevan P, "Anarchism, Mutual Aid, and Self-Organization: From the George Floyd Uprising to India's Farmer Rebellion," *It's Going Down*, March 5, 2021, https://itsgoingdown.org/anarchism-mutual-aid-and-self-organization-from-the-george-floyd-uprising-to-indias-farmer-rebellion.
42. Randy Shaw, *The Activist's Handbook: A Primer for the 1990s and Beyond* (Berkeley, CA: University of California Press, 1996).
43. CrimethInc, "The June 2013 Uprisings in Brazil. Part I," June 27, 2013, https://crimethinc.com/2013/07/27/the-june-2013-uprisings-in-brazil-part-1; CrimethInc, "Chile: Looking Back on a Year of Uprising," October 15, 2020, https://crimethinc.com/2020/10/15/chile-looking-back-on-a-year-of-uprising-what-makes-revolt-spread-and-what-hinders-it.
44. Kali Akuno, "Tales from the Frontlines: Building a People-Led Just Transition in Jackson, Mississippi," in Morena, Krause, and Stevis, *Just Transitions*. p. 155.
45. *Ibid.*, p. 150.
46. *Ibid.*, p. 151.
47. *Ibid.*, p. 163.
48. *Ibid.*, p. 164.
49. Personal communication, April 2021.
50. Diana Cardona and Maria Conill Hernández, "Vallcarca, fer un barri des de la lluita," *La Directa*, February 9, 2021, https://directa.cat/vallcarca-fer-un-barri-des-de-la-lluita.
51. Personal communication, February 2021.
52. Ioanna Manoussaki-Adamopoulou and Alex King, "Inside Exarcheia: The Self-Governing Community Athens Police Want Rid of," *The Guardian*, August 26, 2019, www.theguardian.com/cities/2019/aug/26/athens-police-poised-to-evict-refugees-from-squatted-housing-projects.
53. Molly Crabapple, "This Refugee Squat Represents the Best and Worst of Humanity," *The Guardian*, June 23, 2017, www.theguardian.com/commentisfree/2017/jun/23/refugee-squat-city-plaza-greece-best-worst-humanity.

54. Patrick O. Strickland, "Greek Squatters Transformed a Deserted Hotel Into a Sanctuary for Refugees. Now, They Face Eviction," June 20, 2017, https://inthesetimes.com/article/a-four-star-response-to-the-refugee-crisis-squat-hotel-athens.
55. Theodoros Karyotis, "Criminalizing Solidarity: Syriza's War on the Movements," July 31, 2016, https://roarmag.org/essays/criminalizing-solidarity-movement-refugees-greece.
56. Abahlali baseMjondolo, "Guide for NGOs, Academics, Activists and Churches Seeking a Relationship with the Movement," May 2007, http://abahlali.org/node/1391.
57. Jane Battersby, "Urban Agriculture and Race in South Africa," in Rachel Slocum and Arun Saldhana (eds.), *Geographies of Race and Food Fields, Bodies, Markets* (Abingdon-on-Thames, Routledge, 2013), p. 124. Battersby is quoting P. Burger, J. P. Geldenhuys, J. Cloete, et al., "Assessing the Role of Urban Agriculture in Addressing Poverty in South Africa," GDN Working Paper Series no. 28 (New Delhi: Global Development Network, 2009), p. 19.
58. Frantz Fanon, *Black Skin, White Mask* (London: Pluto Press, 1986 [1967]).
59. Faranak Miraftab, "Insurgent Planning: Situating Radical Planning in the Global South," *Planning Theory* 8(1): 32–50 (2009), p. 44. I drew the quote from Battersby, "Urban Agriculture and Race in South Africa."

4. VERSATILE STRATEGIES

1. Buck, *After Geoengineering*, p. 206, referring to Naomi Klein, *This Changes Everything: Capitalism vs. the Climate* (New York: Simon & Schuster, 2014).
2. For a brilliant history relating to this descriptor, Peter Linebaugh and Marcus Rediker, *The Many Headed Hydra: Sailors, Slaves, Commoners, and the Hidden History of the Revolutionary Atlantic* (Boston, MA: Beacon Press, 2000).
3. Peter M. Rosset and Lia Pinheiro Barbosa, "Autonomía y los movimientos sociales del campo en América Latina: un debate urgente," *Aposta: Revista de Ciencias Sociales* no. 89 (2021).
4. George Katsiaficas, *The Subversion of Politics: European Autonomous Social Movements and the Decolonization of Everyday Life* (Oakland, CA: AK Press, 2006).
5. Naomi Klein, *On Fire: The Burning Case for a Green New Deal* (London: Penguin Books, 2019), p. 238.
6. Javier Ochoa, "DGA lo confirma: Lago Lleu Lleu tiene una de las aguas más limpias de Sudamérica," *Diario Concepción*, December 27, 2017, www.diarioconcepcion.cl/economia-y-negocios/2017/12/27/dga-lo-confirma-lago-lleu-lleu-tiene-una-de-las-aguas-mas-limpias-de-sudamerica.html.
7. John Severino, "With Land, Without the State: Anarchy in Wallmapu," October 18, 2010, https://theanarchistlibrary.org/library/john-severino-with-land-without-the-state-anarchy-in-wallmapu. I also recommend the archives of the Paismapuche.org website.

8. Gelderloos, *Failure of Nonviolence*; Gelderloos, "Riots and Remembrance on the Streets of Barcelona."
9. Peter Gelderloos, "Debunking the Myths around Nonviolent Resistance," August 22, 2020, https://roarmag.org/essays/chenoweth-stephan-nonviolence-myth.
10. Natasha Hakimi Zapata, "Extinction Rebellion's Long Overdue Reckoning with Race," *The Nation*, October 5, 2020, www.thenation.com/article/politics/extinction-rebellion-climate-race.
11. Anonymous, *The Unquiet Dead: Anarchism, Fascism, and Mythology*, anonymous publication and distribution, 2019.
12. Peter Gelderloos, "Diagnostic of the Future: Between the Crisis of Democracy and the Crisis of Capitalism, a Forecast," November 5, 2018, https://crimethinc.com/2018/11/05/diagnostic-of-the-future-between-the-crisis-of-democracy-and-the-crisis-of-capitalism-a-forecast.
13. Cory Robin, *The Reactionary Mind: Conservatism from Edmund Burke to Donald Trump* (Oxford: Oxford University Press, 2018).
14. Andrew Kordik, "Welcome to the Age of Modern Monetary Theory: It's Turning Conventional Economics Upside Down," *Salon*, March 20, 2021, www.salon.com/2021/03/20/welcome-to-the-age-of-modern-monetary-theory-its-turning-conventional-economics-upside-down.
15. Mark Paul, "The Economic Case for the Green New Deal," *Forbes*, February 20, 2019, www.forbes.com/sites/washingtonbytes/2019/02/20/the-economic-case-for-the-green-new-deal.
16. Arrighi, *The Long Twentieth Century*, p. 287.
17. Clint Smith, "Stories of Slavery, From Those Who Survived It," *The Atlantic*, March 2021, www.theatlantic.com/magazine/archive/2021/03/federal-writers-project/617790.
18. Klein, *On Fire*, p. 262.
19. Asad Rehman, "The 'Green New Deal' Supported by Ocasio-Cortez and Corbyn is Just a New Form of Colonialism," *Independent*, May 4, 2019, www.independent.co.uk/voices/green-new-deal-alexandria-ocasio-cortez-corbyn-colonialism-climate-change-a8899876.html.
20. Rajni Kothari, *Rethinking Development: In Search of Humane Alternatives* (Delhi: Ajanta, 1998), p. 143.
21. Phoebe Holmes, "The Political Economy of Biofuels in Mozambique," March 2021, https://africasacountry.com/2021/03/the-political-economy-of-biofuels-in-mozambique.
22. Naomi Klein (*On Fire*) has the right rhetoric but is remarkably short on details. Cooperation Jackson (Akuno, "Tales from the Frontlines,") provides a solid social and racial justice focus, but they go far beyond Representative Ocasio-Cortez and Senator Markey's GND proposal, itself rejected as far too left-wing by the Democratic Party. They do not seem to explain how to bridge this gap and pressure the Democratic Party into accepting a program it is inimical to. Noam Chomsky and Robert Pollin offer a fully technocratic proposal, very sensitive to financial interests, without bothering to grasp the human dimensions, pay attention to social movements, or perceive the

many criticisms relating to racism, coloniality, gender, and all the other interconnected oppressions and catastrophes (*Climate Crisis and the Global Green New Deal: The Political Economy of Saving the Planet*, London: Verso, 2020).

23. Andreas Malm, *Corona, Climate, Chronic Emergency: War Communism in the Twenty-First Century* (London: Verso, 2020).
24. CrimethInc, "1919: When the Bolsheviks Turned on the Workers: Looking Back on the Putilov and Astrakhan Strikes, One Hundred Years Later," March 12, 2019, https://crimethinc.com/2019/03/12/when-the-bolsheviks-turned-on-the-workers-looking-back-on-the-putilov-and-astrakhan-strikes-one-hundred-years-later; Isaac Deutscher, "Chapter II: Trade Unions and the Revolution," *Soviet Trade Unions: Their Place in Soviet Labour Policy*, 1950, www.marxists.org/archive/deutscher/1950/soviet-trade-unions/ch02.htm.
25. Peter Gelderloos, "Charting Revolt: Resisting the Tendency towards Reactionary Sociology," January 19, 2021, https://anarchistnews.org/content/charting-revolt-resisting-tendency-towards-reactionary-sociology.
26. Ashley Dawson, *Extreme Cities* (London: Verso, 2017).
27. John Severino, "Evo's Highway: Development in Socialist South America," December 22, 2010, https://theanarchistlibrary.org/library/john-severino-evo-s-highway.
28. Bill Weinberg, "Indigenous Anarchist Critique of Bolivia's 'Indigenous State': Interview with Silvia Rivera Cusicanqui," September 3, 2014, https://upsidedownworld.org/archives/bolivia/indigenous-anarchist-critique-of-bolivias-indigenous-state-interview-with-silvia-rivera-cusicanqui.
29. Dan Sabbagh, "MI5 Involvement in Drone Project Revealed in Paperwork Slip-up," *The Guardian*, March 6, 2021, www.theguardian.com/uk-news/2021/mar/06/mi5-involvement-in-drone-project-revealed-in-paperwork-slip-up.
30. Josep Gardenyes, "23 Theses Concerning Revolt," *Return Fire* 6(1) (2020 [2011]).
31. Dunlap, "Renewing Destruction," p. 138.

5. A TRULY DIFFERENT FUTURE

1. Dawson, *Extreme Cities* is a good book for imagining the tribulations of coastal cities in the coming decades, though its focus is the United States.
2. Dolors Álvarez, "Seat reanuda esta noche la fabricación de respiradores tras recibir autorización de Sanidad," *La Vanguardia*, April 2, 2020, www.lavanguardia.com/economia/20200402/48272122466/seat-respiradores.html
3. As I write, there are already a number of organizations like Open Arms that run boats to help people survive the Mediterranean passage; another, Women on Waves, that perform abortions in international waters for people in countries like Poland; and a Cofradía de navegantes anarquistas

or "anarchist sailors' guild" along the Iberian coast connected to similar groups around the world.

4. Peter Kropotkin, *Fields, Factories, and Workshops* (Boston, MA: Houghton-Mifflin, 1889).
5. Buck, *After Geoengineering*, p. 78.
6. Institut Català d'Energia, "Balanç d'energia elèctrica de Catalunya," 2021, http://icaen.gencat.cat/ca/energia/estadistiques/resultats/anuals/balanc_energia. I slightly changed the data provided here to reflect projected growth of renewables.
7. This technique is discussed for lay readers in Buck, *After Geoengineering*, pp. 104–105.
8. Michelle Toh, "Rio Tinto Blew Up a Sacred Site in Australia. The CEO Left but Still Got a Huge Payout," *CNN*, February 22, 2021, https://edition.cnn.com/2021/02/22/business/rio-tinto-annual-report-2020-intl-hnk/index.html.
9. Ward Churchill, *A Little Matter of Genocide: Holocaust and Denial in the Americas 1492 to Present* (San Francisco, CA: City Lights Publishers, 2001).
10. M. De Vabres, "Judgment: Doenitz," The Avalon Project, https://avalon.law.yale.edu/imt/juddoeni.asp.
11. Eric Boehlert, "How the Iraq War Still Haunts New York Times," *Media Matters*, July 1, 2014, www.mediamatters.org/new-york-times/how-iraq-war-still-haunts-new-york-times. I would add the observation that on a day-by-day basis during the lead-up to the war, *Democracy Now* was broadcasting well-researched rebuttals to the NYT's single source allegations.
12. Claire Suddath, "Who Are the Gypsies, and Why Is France Deporting Them?" *Time*, August 26, 2010, http://content.time.com/time/world/article/0,8599,2013917,00.html [I would editorialize that the mere title of this piece illustrates the point I am trying to make about unequal memory and unequal protections]; Carol Silverman, "Persecution and Politicization: Roma (Gypsies) of Eastern Europe," *Cultural Survival*, June 1995.
13. The Truth and Reconciliation Commission of South Africa Report, vol. 7, 2002.
14. Peter Storey, "A Different Kind of Justice: Truth and Reconciliation in South Africa," *The Christian Century*, September 10–17, 1997.
15. Brandon Hamber, Traggy Maepa, Tlhoki Mofokeng, et al., "Survivors' Perceptions of the Truth and Reconciliation Commission and Suggestions for Final Report," The Centre for the Study of Violence and Reconciliation & the Khulumani Support Group, 1998, archived at https://web.archive.org/web/20060925181412/www.csvr.org.za/papers/papkhul.htm.
16. "Guatemala indemniza a los familiares de las víctimas del asalto a la Embajada de España," *El País*, January 30, 2007, https://elpais.com/internacional/2007/01/30/actualidad/1170111602_850215.html
17. Michael Gould-Wartofsky and Kelly Lee, "Guatemala's Dirty War," *The Nation*, May 15, 2009, www.thenation.com/article/archive/guatemalas-dirty-war.

18. Tawinikay, "Autonomously and With Conviction: A Métis Refusal of State-Led Reconciliation," October 22, 2018, https://north-shore.info/2018/10/22/autonomously-and-with-conviction-a-metis-refusal-of-state-led-reconciliation.
19. *Ibid.*
20. Indigenous Action Media, "Accomplices Not Allies: Abolishing the Ally Industrial Complex, An Indigenous Perspective," May 4, 2014, www.indigenousaction.org/accomplices-not-allies-abolishing-the-ally-industrial-complex; M., "From Charity to Solidarity: A Critique of Ally Politics," in Cindy Milstein (ed.), *Taking Sides: Revolutionary Solidarity and the Poverty of Liberalism* (Oakland, CA: AK Press, 2015).
21. Tawinikay, "Autonomously and With Conviction."
22. Ilan Pappe, *The Ethnic Cleansing of Palestine* (Oxford: One World Publications, 2006), pp. 52, 107, 109; Canadian Utopias Project, "Sointula, British Columbia," https://canadianutopiasproject.ca/settlements/sointula-bc.

Index